Real Life After Life

About the author:

The natural scientist Dipl.-Math. Klaus-Dieter Sedlacek, born in 1948, studied mathematics, computer science and physics in Stuttgart. After twenty-five years of professional experience in his own company, he now devotes himself to his private research projects and publishes the results in a generally understandable form. He is also the editor of several book series, including the series "Wissenschaftliche Bibliothek" and "Wissen gemeinverständlich".

Klaus-Dieter Sedlacek

Real Life After Life

The liberation of consciousness from the shackles of time

Wissen gemeinverständlich Bd 03
English edition

Bibliografische Information der Deutschen Nationalbibliothek: Die Deutsche Nationalbibliothek verzeichnet diese Publikation in der Deutschen Nationalbibliografie; detaillierte bibliografische Daten sind im Internet über dnb.dnb.de abrufbar.

© 2020 Klaus-Dieter Sedlacek
https://toppbook.de

Cover composition: Sedlacek

Production and publishing
Herstellung und Verlag: BoD - Books on Demand, Norderstedt

ISBN: 978-3-7504-0935-4

Table of contents

1. Preface 7
2. Introduction 9
3. The Life 11
   - Metabolism 17
   - Regeneration 21
   - Mind and soul 31
   - The body and soul problem 34
   - Summary of the chapter 36
4. From awareness to information 38
   - Why animals decide about their behaviour 41
   - Regulation in humans 47
   - Consciousness as a manifestation in its own right 54
   - Can ASW be explained scientifically? 80
   - Is there a real afterlife? 83
   - How information containing energy is generated 88
   - Where is the generated information? 92
   - Summary of the chapter 94
5. Is there life after the end of time? 96
   - The survival of human consciousness 103
   - Analogies and laws of nature 114
   - The organization of life after life 124
   - Summary of the chapter 128
6. Literature 132
7. Keyword index 137

1. Preface

What can natural science contribute to a topic that is otherwise occupied by philosophy and the various faiths?

I think a lot. For in natural science there are the extremely strange phenomena of quantum physics, which despite their oddity have the potential to shed light on those areas. However, it is about things where the established natural scientist usually simply turns away in order not to expose himself and his department to the danger of possible criticism.

But I don't believe that as a natural scientist you get dirty when you take up a topic that is close to the hearts of many people. These are people who may not be able to cope with what they are offered as wisdom on Sunday morning before the news. I have nothing against wisdom, I'm sure it is important for many people. It is just that some wisdom has a depressive character, in contrast to the claim of "good news".

So I think it is time to not only claim the good news, but to prove it. And evidence is rather the field of natural science than philosophy or good advice.

I do not want to claim that I have proved everything that is written in the following chapters, but I have chosen the path of reason, the path of comprehensible thinking, the scientific path as far as possible.

However, this path is also not without its stumbling blocks. First of all, there is the problem that the natural sciences have not defined consciousness; behavioural psychologists and philosophers have so far done so. But I needed a scientific definition. So I have created one myself, which is derived from what behavioral psychologists consider to be good and correct.

And then there is a physical phenomenon which Albert Einstein scornfully called "spooky long-distance effect". Today's physicists use terms such as "notlocal" or "quantum entanglement" instead. The spooky aspect of the phenomenon has disappeared from the vocabulary, but it exists, as experiments have confirmed.

When the word "notlocal" is mentioned, however, physicists unpack their rubber gloves so that they only come into superficial contact with them. They no longer explain what "notlocal" actually means, but rather seal themselves off. Everything that belongs to space and time of our known world is "local". The physicists still agree on that. Then "notlocal" can only mean It is something that does not belong to space and time of the world known to us.

I saw myself forced to draw this and similar conclusions myself. In addition to my sometimes somewhat sober, albeit generally understandable explanations, I am fortunate to owe many formulations to Prof. Dr. Dennert, who wrote them down some time ago, so that the text as a whole has come alive.

So I can only hope that you, dear reader, will actually find the contents of this book a pleasant message.

Stuttgart, spring 2016.

Klaus-Dieter Sedlacek

2. Introduction

The question of the temporal end of our existence, which is of great importance to us human beings, and the rational answer to it should be our task in this book, but not the answer that faith seeks.

I am using the "we" here because I want to take you with me on the path of answering, which I would like to take together with you.

So in the following we want to stay on the ground of science, and there we have to examine quite soberly how far from the facts and from the point of view of logical thinking we can reveal the mystery of the temporal end and answer the question: Is there life after life?

If we have relentlessly followed this path to the end, then every reader may decide whether he wants to go another way, the way of the heart, with a clear conscience. Perhaps that this then still illuminates in a peculiar way many a stretch of the path of reason, which by its nature had to remain dark.

We can also call the path of rational thinking the critical-scientific one. But then it consists in the fact that we examine all phenomena that are connected with the question in question, that is, for us with the question of the end of time, without prejudice, and extract from them what is generally valid, whereby we must never deviate from logically clear concepts and conclusions.

So let us never try to leave this path in these investigations, only then will we keep firm ground under our feet until the end.

Now, it would seem extremely difficult from the outset to find a suitable beginning for our journey, because after all it is something which we can safely call a "secret".

But there is a way, and a very promising way.

Our basic question, of course, must be: "What is the temporal end?" But we will also be able to reach our goal by asking first of all for the possible opposite of the temporal end.

When we can understand this then we will also be able to unravel the mystery of the temporal end from him in the end.

We see life as the opposite of the temporal end, and since we are all in life, since life is our very own possession, we will have something to say about it.

Of course, we do not want to hide from the outset that here too we are faced with a serious problem, the mystery of life. This much alone makes it clear that it must be the less difficult problem for us, precisely because we ourselves are in it, and that we can best understand the mystery of the temporal end from the problem of life. Therefore, this will be our first concern.

3. The Life

What is life[1]? We ask this question first.

Life is a natural phenomenon, it belongs to nature, - about this there is not the slightest doubt at first. But if this is the case, then we find ourselves on the ground of natural science with the question of life. And she must be able to help us answer the question.

A look into nature and at the bodies and figures that compose it immediately shows us that we have to distinguish between these two large groups: Unanimated and living natural bodies. To those belong stones, rocks, earth, water, air, - to those plants, animals, people. To the unbiased observer this dichotomy seems self-evident.

But is it justified and what does it mean?

From the so-called monism[2] one has denied that one has to make such a distinction between the inanimate and the living and has claimed that the living is only a special form of the inanimate. The starting point was the philosophical endeavour to attribute everything real to a single principle, an endeavour which, however justified it may be in itself, must not influence us in such a way that we view nature with bias. - So we want to keep ourselves free of it on our way to the knowledge of life and the temporal end.

Let us first try to dissect the inanimate.

The inanimate consists of material substances of various kinds and is characterized by certain proper-

---

[1] Detailed treatment of the topic in the book Dennert *"Das Geheimnis des Lebens"*, Halle a. S., N. Mühlmann.
[2] **Monism** is the viewpoint according to which the physical and the spiritual are different, equal manifestations or sides of one and the same real thing.

ties. If we observe the behaviour of this matter, we see that it is subject to constant change. Yes, on closer inspection we find that all events in nature are basically nothing more than a continuous change of matter, but that this change is ultimately based on the interaction of the various material substances of the world.

The water of the streams and seas becomes vaporous through interaction with the temperature conditions of the air and rises into the air as mist, conglomerates here in clouds and becomes solid through cold air currents, falls down as snow, which melts on the warm ground and becomes water again, etc.

The hard rock becomes crumbly and chemically altered by the interaction with the factors of its environment (air and water), finally it disintegrates into arable earth, the material substances of which are transferred to the plant through the life process of the plant, and from the plant often into animals and humans. When their body later decays, its substances are again decomposed and are again transferred to the substances of the earth.

In the same way, we could dissect what is happening in the whole world of material matter, everywhere we would discover sequences of such changes and interactions, work performances or energies, as we call it scientifically. If the substance itself changes during the processes in question, e.g. iron changes into brown rust during rusting, then this is a chemical work performance, but if only the state changes, but not the substance itself, e.g. when the iron is magnetized, when the water becomes steamy, etc., then we speak of physical energies.

It is a very significant result of modern natural science that all natural phenomena occur in material substances and that all changes in substances can be traced back to chemical and physical energies.

It is irrelevant for our present purpose to pursue this further in detail, but the following is extremely important: Energy or work performance can always be measured.

As far as measurable, we are completely in the field of inanimate matter and its energies and thus in the field of chemistry and physics.

And another one! If we follow the transformations of energies and matter more closely, we come across a highly significant fundamental law of inanimate nature in classical physics[3], which is
During all transformations of material substances in a closed system, the mass of matter remains constant and the sum of energies remains unchanged.

If we follow any changes in certain substances, we make the surprising discovery that neither substance nor energy is lost or newly created. Substance and energy are changeable, but indestructible.

If, for example, any substance burns, it combines with the oxygen in the air as a result of chemical energy. Precise weight measurements have shown that the total mass of all substances involved in the process is the same before and after, no matter how profoundly they have been changed.

---

[3] In quantum and particle physics, the situation is a little different. There are transformations between mass and energy. There the energies remain unchanged in the total.

So if you weigh the material to be burned together with the oxygen available to it and then weigh all the combustion products again after combustion, we get the same number. Thus, nothing was lost in substance (in mass) during combustion.

And it is the same in all other cases, including physical processes, e.g. when movement is converted into electrical energy and this is converted into light, etc.

So these energies can always be measured, and therefore, for example, electric current and electric light can be sold like a commodity, which would be completely impossible if there was no measurability.

With the measurability of energies we are, as already mentioned, in the area of inanimate matter.

Is there now another characteristic of these phenomena of the substance? The continued observation of world events, however, reveals another very important one. It has been shown that the same causes have the same effects. Where the same conditions meet, the success is always the same.

For example: Wherever in the world hydrogen burns, water is created, wherever a corresponding temperature is applied to water, it becomes solid or vaporous, etc. And these processes also always take place according to very specific measurable conditions. Because the same phenomena also always have the same effects, therefore world events give us the impression of necessity, of regularity, and this regularity is the further unmistakable characteristic of phenomena in the field of inanimate matter.

What about the living now?

First of all, it must be said that life too is always bound to matter. We don't know it any other way. And this matter has the same chemical and physical properties as in inanimate nature. Of course, it is after all a special material substance.

All living beings are composed of cells, but cells are lumps of so-called protoplasm with a cell nucleus. Chemically, the protoplasm consists mainly of proteins. These now do not exist outside of living beings in nature. Wherever we find them, we can conclude with certainty that they originate from living beings. But these proteins are made up of the same basic materials that make up the substances of the inanimate world.

But protoplasm is not completely synonymous with protein. Its peculiarity is that it is organized, that is, it has a special peculiar construction. In order to understand this, we need to look at something else first.

A drop of water can undergo the most diverse changes, become steamy, become solid, or even be broken down by the electric current into its basic materials, hydrogen and oxygen. It is then always possible to reconstitute it, even from the basic materials. So he is indestructible in some ways. And it is the same with all other inanimate natural bodies.

In contrast to this life is a temporary phenomenon and every living being without exception approaches its temporal end and when it is dead then according to the present state of science it is impossible to recall it back to the living state. A borderline case of this rule are the experiments of the American biochemist Craig Venter, who was the first to succeed in

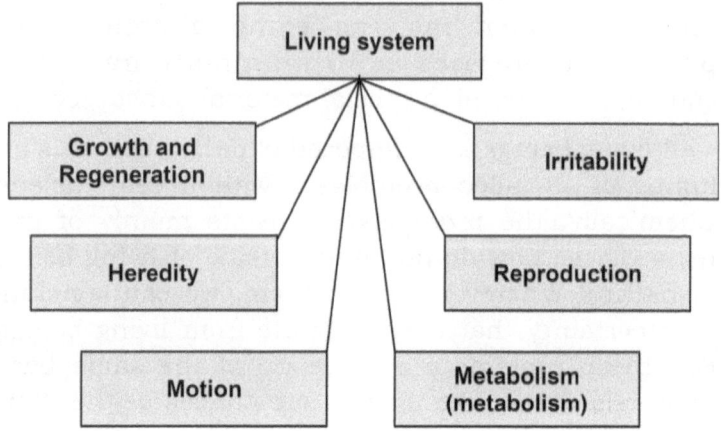

Fig. 3.1: Characteristics of a living system. Graphic: Sedlacek

producing a genetic material himself and implanting it into a cell, thus creating a viable bacterium.[4]

Nevertheless, the certainty that life is something other than the dead state arises. Every death proves it with compelling certainty.

It is now also the case that every living creature is constantly exposed to the danger of death. In order to escape the end of its temporal existence, it must constantly carry out a series of actions, i.e. processes, which have the very purpose of sustaining life: It must move, it must feed itself, it must respond to harmful stimuli appropriately and protect itself. And since every living being must die one day, it must reproduce and produce offspring beforehand, if all life on earth is not to disappear in a short time (see Fig. 3.1).

---

[4] https://de.wikipedia.org/wiki/Craig_Venter

Incidentally, a process is defined as the totality of interacting procedures in a system by which matter, energy or information is transformed, transported or stored.

So these are the processes of life, and we cannot discover anything like this in any inanimate natural body. Wherever one believes it, it happens under revaluations and reinterpretations of fixed terms, which are made for the sake of other philosophical views.

So the processes of life always take place in such a way that they serve to preserve life, and that is why they are called purposeful. And you will not find such practicality anywhere in inanimate nature. There is no point at all in opposing this word, as is actually happening from some quarters. Because then you would have to find another word for it, such as "useful" or "needy".

The result, however, always remains the same: that there is no such thing in inanimate nature. That here, then, there is a fundamental opposition between the living and the inanimate.

In order to understand all this even more clearly, we will now take a closer look at two life processes, metabolism and regeneration.

Metabolism

Substance is consumed in all life phenomena. The time limit would soon come to an end if this substance were not replaced. So here there is an undeniable need of living beings, and this need is fulfilled exactly as the preservation of life demands it.

A need is defined as the tendency to pursue a goal. Inclination indicates that it is not the rigid, unalterable pursuit of a goal. Conflicting objectives may need to be resolved when two different objectives are contradictory. Then a decision must be made. Further down we come back to the concept of need when we take a closer look at consciousness.

To pursue a need special devices, tools, are needed on the body of living beings, and these tools are called organs and the construction of such organs is called organization or system.

This is not only true with regard to nutrition, but also with regard to all other life phenomena.

That is why every living being shows organization, and this goes right down to its smallest components: the cell, even the protoplasm, is organized in this sense.

The living beings are differently perfect, the simplest ones consist only of single cells; but then one can observe life processes, e.g. nutrition, and an organization serving for it.

Nutrition in the whole of the living world, in the simplest and most complex living beings, consists in the absorption of different substances from the environment by certain organs and their processing in other organs into the substances from which the living being is built, and always into the substances that are needed here or there, according to need. This so-called metabolism takes place under constant disturbance of the chemical balance.

There is not a single living creature where it would not be so, and there is not a single inanimate natural body where something even remotely similar could be observed. In inanimate nature, the chemical processes are rather constantly striving towards chemical equilibrium.

But when a previously living being has died, has become addicted to death, then it immediately behaves with regard to its chemistry just like an inanimate natural body, then no power of the world can again cause the nutrition process and metabolism in it: Everything now strives for chemical balance.

That transformation of the ingested food into the substances of one's own body, primarily into proteins, is now in itself quite chemical in nature and in this respect does not differ fundamentally from the processes in inanimate nature. It is not the nature of the processes that is peculiar to life, but the fact that they always take place as required by the need to preserve life, i.e. in relation to the organism.

This will become very clear when we consider a particular example.

In our liver cell, the size of which is about a thousandth of the size of a pinhead button, very different, often almost opposite chemical processes take place next to each other, depending on what the preservation of life requires. Ten or more such events have been counted. For example, sugar is produced in the liver cell from a starch-like substance, glycogen, to provide the blood with the necessary amount of such a substance. However, if the blood contains too much sugar, so that life is in danger, then in the same liver cell, this sugar is converted back into glycogen.

Franz Hofmeister says about it[5]: *"What distinguishes the whole process here is the amazing simplicity and practicality of the means used and the resulting saving of space and energy.*

---
[5] *Die chemische Organisation der Zelle.* Braunschweig 1901, p. 13.

The protoplasm of the cell carries out these chemical processes with the help of very special, peculiar substances called enzymes, which have the highly striking property that they only cause chemical processes by touching them or by making changes themselves very easily, but that they then just as easily regress in order to act again. In any case, very little of such enzymes is needed to trigger one and the same chemical process all the time. They will be highly useful for chemical processes in the extremely small space of a cell.

It is therefore very remarkable that it is precisely in the cells of living beings that such enzymes are used in their nutritional processes, whereas they are not known in inanimate nature.

The microscopically small liver cell of which we spoke earlier contains no less than ten such different enzymes with different areas of activity. They cause and regulate the important chemical processes in this cell that have already been touched.

It is no different with the simplest living beings consisting of only one cell. In the yeast fungi, which consist of microscopically small single cells, no less than four different enzymes were found: one, diastase, for the conversion of starch into sugar, a second and third (invertase and maltase), which cause conversions of different types of sugar, and a fourth (zymase), whose field of action is fermentation, whereby alcohol is produced from sugar. And also here all these processes, which are in themselves quite chemically based, always take place according to the life requirements of the yeast fungus.

Regeneration

The second life phenomenon that we want to use here to confirm what has been said is regeneration. -- We have already said that living beings must protect themselves against harm; their useful devices and abilities in this direction are very great, and many are precautionary for possible future dangers. However, a special chapter offers the question: How does the organism help itself when direct damage to the body has already occurred?

Well, in this case he has the ability to immediately repair and heal this damage, i.e. wounds, and even to replace lost organs in some animals. This is called regeneration. It can be said that the less a living being is developed as a whole, the more it is gifted with the ability to regenerate.

If you try to catch a lizard and grab it by the tail, it will break off very easily. This is an important means of protection for the animal, because while you hold the still wriggling tail in your hand, the animal itself has long since escaped. From the wound, however, a new tail develops bit by bit.

Frogs are able to rebuild legs that have been cut off; many worms, including earthworms, regenerate to such an extent that each piece cut off grows into a complete worm.

In a very peculiar way, the oystercatchers have experienced the regenerative ability of an animal to their detriment.

The well-known starfish, which has five star-shaped arms, is an enemy of the oysters. He knows how - we still don't quite know how - to separate the two shells of an oyster and then eat the animal.

The oystercatchers are therefore, as is understandable, quite bad at talking about starfish, they catch them where they can and try to destroy them. First they did this by cutting up the animals and then throwing them carelessly into the sea. However, they had made the calculation without the host, in this case without the wonderful regenerative capacity of starfish. After some time they discovered that they now even had five times as many starfish in their area as before. But the new starfish all had one striking peculiarity: they had one long arm and four short arms.

The regeneration explains the riddle: Four arms were newly formed from the wound of one cut off arm, so that now a complete starfish was formed from the one arm again.

Of course, the oystercatchers had learned from the damage and did not throw the slashed starfish back into the sea.

Many plants form a bud from cut leaves that are kept in moist soil, from which a new complete plant is then formed. - In some trees, which are planted upside down in soil, a new crown is obtained from the root part, while the old crown takes root in the soil.

These examples can be multiplied at will. The higher beings also possess the ability to regenerate, even man, although to a much lesser extent, since he does not really need it because of his other abilities. - But the ability to heal wounds is also a regeneration, and how far the latter goes, for example in the case of fractures and splintering of bones, is often experienced by the doctors on our wounded patients to their own astonishment.

It is particularly noteworthy that even whole eyes can be newly formed in cancers and amphibians. However, the most wonderful phenomenon in this direction is the following: They took away a cancer's eye and then kept it in the dark. And lo and behold, the wound did not give rise to a new eye, which he could not have used in the new circumstances after all, but - a feeler, i.e. an organ of touch, which in its entirety is calculated for orientation in darkness.

Especially from this last example it results with striking logical proof that regeneration, and thus also every other life process takes place according to need, thus does not run in well-worn, necessary tracks, but with a certain freedom, but always -- expediently related to the organism.

In order to see through these circumstances quite clearly, the following must now be emphasized: Living beings also consist of materials and work is carried out on these materials, so here too we observe the twin team of material and energy. - And in their activity, the same laws result as those that have been researched in chemistry and physics in the field of the inanimate.

> *The laws of conservation of mass and energy, which we have come to know as the basic laws of all inanimate nature, also prevail in the matter of living beings.*

And finally: In him, too, everything happens according to the law of cause and effect; therefore, it also says here that equal causes are followed by equal effects. However, it is the case that this law does not apply at the atomic level to individual atoms or elementary particles such as electrons in this generality. Because there, due to the principally restrictive laws of nature, we can only predict the probability of later observations. In the scale of a whole system, as we have with one cell, this restriction does not play a role.

All are therefore characteristics not only of inanimate matter, but of matter in general. If 150 years ago at the time of the so-called old vitalism this was denied and one believed in a special law of a living substance as expressions of a special "vital force" on this special substance, this was a mistake, as nature research has meanwhile revealed.

On the other hand, it is a much worse fallacy to conclude that the life processes are completely absorbed in the chemical-physical processor that there is no fundamental difference between the inanimate and the living. This would be quite the same as saying that this wall here was made with paint with a brush, and since the Raphael Madonna is also made with paint with a brush, there is no fundamental difference between this wall and the Raphael Madonna.

On the contrary, our investigation has shown that, despite all the laws of chemical-physicalevents in the processes of life, a certain something is added to these processes of life, which we can never observe in inanimate natural bodies. This is the following circumstance: The chemical-physical and purpose-related phenomena always occur in such a direction that they serve to preserve life.

In the inanimate nature, the energies work on the materials always and only according to the one point of view which is necessarily given by the existing factors: the water evaporates or becomes solid, depending on the existing conditions, one can never ever speak of a "purpose" for the behaviour of the water; the wind blows from the east or from the west, as a weak whisper or as a strong storm, always necessary because it is lawful, never and never in one case like this, but in another case like this, because it would have the purpose to satisfy a specific need. To talk about it would be nonsense.

But with the living being, everything that happens is dictated by the purpose of satisfying existing needs. How it happens is chemicallyand physically determined and necessary; -- but that or whether it happens is regulated from a special point of view, namely from the point of view of need. Here the question is always decisive whether it is useful for the preservation of life.

In other words, in life we find ourselves in a field in which necessity and freedom are linked in a most remarkable way, without one noticing anything of the supposed dichotomy of these terms.

Life is therefore not an opposition of the inanimate, but a superordination, a mastery of the dead. The inanimate matter and its energies are not suspended in life and through life, as would be the case if they were opposites, but are controlled, directed in a certain direction, each time determined by the need to preserve life, that is, by expediency. Thus the relationship between the inanimate and the living is completely clear. After all we said, we can say it now:

In living beings, the energetic processes or, in other words, the work of the substances are directed in a purposeful way to satisfy needs.

Nowhere else can you observe such a thing in inanimate matter. So it's not the stuff itself that's alive.

In living beings, therefore, something that gives direction or is conductive is added to the chemical-physical process, i.e. to the substance. That this guiding or guiding force would not be something material or an energy, even if it is a very special, for example vital energy, is completely excluded, because it lacks the main characteristic of energies, the measurability. To talk about measurability of this functional management is complete nonsense. The result may show a lower or higher degree of usefulness; but the line itself is either useful or not useful.

Let's remember our example from the liver cell. We saw that various chemical processes take place in it with the help of the enzymes present in it. This process produces, for example, more or less sugar from glycogen or vice versa. These conversions take place in accordance with those laws of inanimate matter that we have described, and all of them can be measured. The quantity of sugar produced, etc. is determined by the quantities of substances present, etc. Above all, we find that the law of conservation of mass within the cell is respected.

But one circumstance remains: These chemical, per se measurable conversions take place just as the preservation of life demands. How should this fact be measured? And by what? The measurable processes in the cell have all already been attributed to energetic phenomena; this fact that these processes coincide with the need for life is a thing in itself, which is added to everything, but falls outside the economy of available energies.

The processes either coincide with the necessities of life, or they do not. There can be no question of measurability.

Because this circumstance, which causes the expediency of the energetic processes, for which one must necessarily also have an explanatory cause, stands above the energetic processes, we call it a "guiding" principle. In him lies the peculiarity of life, and according to him life is to be regarded as something fundamentally different from the inanimate, from the material.

The described expediency is a pattern that can assume two values, expedient or not. Such a pattern is called information[6]. This information has a certain meaning for the guiding principle, in that it controls the chemical-physical interactions of the substance "expediently". Thus the guiding principle is what we have defined as a process above (p. 17).

Thus, in addition to the energetic manifestation which dominates the material in general, there is a second manifestation in the world which dominates the material of living beings and its energies in particular. This second manifestation consists of controlling processes which control the biology[7] of life and which we want to call in its entirety a **bio-regulatory system** (or in short, a **regulatory system**).

---
[6] **Information** is a pattern of matter or a form of energy that has a certain meaning for a physical process by controlling the way an interaction takes place.

[7] **Biology** is the science of life and living beings.

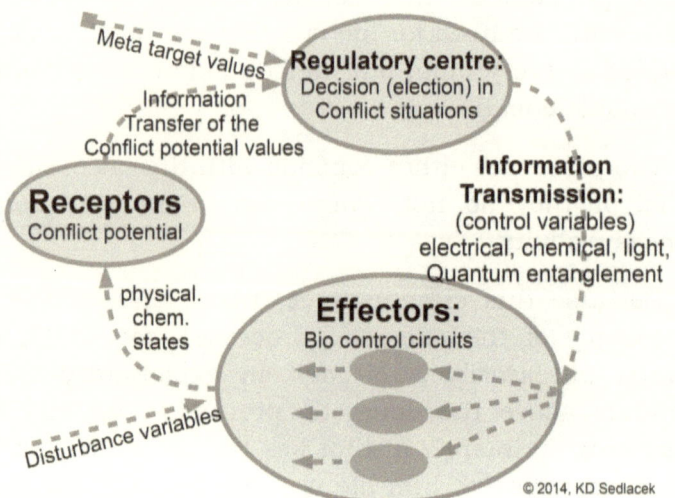

Fig. 3.2: Schematic representation of a biological control loop, the effectors of which in turn are control loops. At the lowest level, the effectors influence metabolic activity, reproductive activity, hormone production or the activity of muscle cells. An important function of the control loop is the processing of the information produced. In the highest level of the control circuits of a system, the control centre is represented by the central nervous system (brain). This is where the information processing that influences and controls the entire system takes place. At the lowest level, the control centre is located in the individual cells, so that specific information processing also takes place within the cells. Graphic: Sedlacek

The humanbio-regulatory system is the totality of the processes that control its biological activities. It contains all regulatory mechanisms or control loops within its organism.

The control loop is a universal principle and always occurs when a goal (normal value) or an equilibrium is aimed for, which does not come about by itself due to disturbing influences or as a result of unstable or indifferent equilibrium systems.

A biological system only remains intact because existing control loops counteract life-threatening interferences. For example, a microbe equipped with flagella only reaches the location of the food substance by chance at most, if its locomotion would not be repeatedly directed to its destination by a control loop despite all disturbing influences.

A control loop needs at least three components to function. Firstly, the controller itself, which ensures that target values are correctly controlled by control information in the form of electrical or biochemicalsignals or light. Target values are either tapped normal values of a biological system or higher target values (meta target values) of the cell group to which the cell belongs. A second important component of a control loop includes the group of effectors, which can change the state of a system. In cells and cell assemblies, these actuators are activated by changes in cell activity and the permeability of the cell membrane. There are four areas where changes can be made:

1. Metabolic activity
2. Reproductive activity
3. secretion activity (e.g. hormone production)
4. Change in activity of the muscle cells

For the controller to be able to output meaningful control information for the effectors at all, it needs information about the current state of the system, i.e. in our case the body. This information is provided by the receptors in biological systems. To make the control loop complete, the exchange of information between the three components is still required. According to our present knowledge, the information exchange takes place electrically (nerve cells), biochemically(e.g. through hormones), through light or through quantum entanglement (see p. 56). Each of

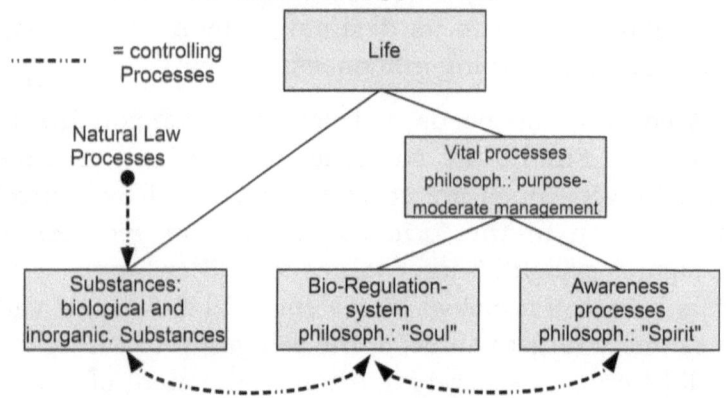

Fig. 3.3: Components of a living system and controlling processes. Graphic: Sedlacek

these forms is simultaneously connected with a transfer of energy.

The bio-regulation system is therefore something guiding and at the same time purposefully effective. It can be seen as a principle of will, a kind of spiritual principle. In technical and scientific terminology, such a mental principle is referred to as an information processing system.

The information-processing system is generally characterized by unconsciousness, because even within us all life phenomena, e.g. nutritional processes, occur without us being aware of them.

The main thing is and remains that this regulatory system cannot be a substance, because it cannot be measured and because it rules over matter and guides it. It has the character of a connection of the material with something mentally abstract, the information.

It is desirable to put the bio-regulatory system in relation to a term that used to dominate many philosophical considerations.

Mind and soul

There are hardly two terms that have such a jumble as "soul" and "spirit". Soon both are used to describe the same, sometimes different things, sometimes they are taken as equally important, sometimes as superior terms, and even then they are allowed to express the most different things.

In this situation, it seems to be the most appropriate way to return to their original meaning with regard to these terms. This is already owed to historical truth, and if, as here, we are dealing with concepts into which further historical development has brought almost endless confusion, then in my view the only possible remedy to get out of this confusion is to go back to the original concept. -- And there the matter is indeed extremely simple.

"Soul" originally referred to the life force that governs living beings as opposed to the inanimate and causes their movements and growth. Later the term changed already in Greek philosophy; already Aristotle distinguished the nourishing soul of plants from the feeling soul of animals and the thinking soul of humans. Apparently two things are already mixed up here, which our further consideration will have to examine.

In any case, however, on the basis of scientific considerations, newer philosophy has shown that it is not justified to make a very sharp distinction between the nourishing and the feeling soul; there is no fundamental difference: We are entitled to summarize both as "soul".

Now, however, what this old term "soul" means is precisely the peculiarity of life in relation to the inanimate, that is, what we are looking for a term for. Even if at that time one may have connected a wrong view with it, in principle it remains the same. And even if we do not want to talk about "vitality" any more, because this could cause misunderstandings, we are nevertheless quite justified in philosophy to use the ancient concept of the "soul" for the peculiarity of life.

According to what we have seen, "soul" is the special one in the material manifestation of life, which guides and regulates the activity of the chemical-physical energies in living beings and thus also an information processing system.

At the beginning of our reflections we said that we wanted to stay on the ground of science. What was meant was natural science. The philosophical concept of the soul is not very helpful for our considerations. To use a scientific term instead, we need only remember how we defined the bio-regulatory system and then compare it with the definition of the soul. We note that both definitions have as a determining characteristic the guiding and regulating of chemical-physical energies. So instead of soul we can use the term bio-regulatory system or just regulatory system, because both are the same.

One more point should be mentioned here. We have said that the manifestation of life, now we say the regulatory system, dominates the body; this becomes particularly apparent when we look at the development of living beings. Here too we find a fundamental difference between "inanimate" and "living".

In contrast to all dead natural bodies, the formation of which takes place merely through the interplay of the energies of the material and its environment, all living beings "develop", i.e. they form from a simple state to greater perfection by gradual formation from within. In this process, the organs emerge from a previously similar and divorced state, the first beginning of which is the egg, i.e. a cell that is organized but does not yet show any signs of the later organs of the being concerned.

The formation of these organs is also a process of the highest expediency, completely dominated by the vital needs of the being in question and quite corresponding to them. And what is particularly remarkable about this process is that it takes place in anticipation and care for the future, in which the institutions will play a role for the first time.

It is obvious that this formation of the organs, this development, also falls entirely under the concept of the manifestation of life, that is, as we now want to say, the bio-regulatory system. The regulatory system has not only the task of maintaining the organism, but also of its formation. It is therefore an organising principle.

The latter, by the way, is completely in line with what we have said about regeneration; because here, too, the activity of the regulatory system is an organising one.

The essential thing in all this is that new formations take place here, that it is a principle that is creatively effective within certain limits, again a very significant difference to the purely energetic manifestation, which never develops any creative activity.

The body and soul problem

Now the question would still be what is the relationship between the regulatory system and the body of the living being. We are thus faced with a problem which has been on the minds of the greatest minds to this day under the heading of the "body and soul problem"; and we cannot and will not undertake to solve this question philosophically, we can only take a general stand on it on a scientific basis.

It is incredibly difficult for the human spirit to think of an information processing system, and so it has also again and again come back to thinking of the soul as a substance, however fine it may be, as an ethereal being etc. The information processing behind this has been hidden. No matter how finely one thinks of the material, it still remains material and as such measurable, which is impossible for the soul, as an information processing system. We must therefore drop the idea of the soul as an ethereal being.

But how does the soul, i.e. the bio-regulation system, affect the body?

This question can be answered quite simple after previous preparations and without etheric body, etheric being or whatever etheric. In our local world, every piece of information and thus every information-processing process always requires a material or energetic information carrier.

A printed e-mail has the paper as information carrier. The television program needs the electromagnetic radio waves as information carrier, which bring the program into our homes, and additionally the television screen as information carrier for the picture, so that we can watch it.

Through the close connection between information and its material carrier, together with the processes that do the actual work, it is possible to control and regulate everything that is intended.

In the television set, for example, specific processes convert the image information contained in the electromagnetic radio waves into the television picture and then ensure that the moving images are displayed on the screen.

In the bio-regulation system many processes are combined, both for the individual cell and for the complete living system, the body. For example, a certain group of processes is responsible for evaluating the genetic information contained in the cells and using this information as a blueprint for enzymes, proteins or other substances.

The processes themselves are mostly located in cellorganelles, which are structurally delimitable areas of a cell with a special function. The processes can also be imagined as small machines constructed from biological materials that are controlled by specific process information. As we can see, there is again a close connection between information and the carrier of the information, namely the biological material.

If you want, you can imagine a separation of information and its carrier. This is possible because information can easily change its carrier. For example, an e-mail, while being sent, has electromagnetic waves as its carrier. Once it has arrived and has been prepared accordingly by processes, it has the screen as its carrier. And when you print it out, it has paper as a carrier.

In the sense of a division one can therefore imagine the soul as information including the process information of the bio-regulation system. One must not forget that information in the local here and now of our world always needs a carrier. But in principle it is possible to transfer information to another medium, as the example with the e-mail shows us. How the conditions in a possibly existing non-local afterlife are represented, we will consider later.

Summary of the chapter

In summary, let us now look back once more.

To get to the bottom of the mystery of the temporal end, we first examined the difference between the inanimate and the living.

We saw: All processes and changes to the materials can be traced back to chemical and physical energies, are always measurable and are strictly legal.

The situation is different, however, in the field of living beings. It is true that they too consist of substances; but these substances alone are characteristic of them, and everything that happens to them is done according to the same energies that are active in inanimate matter.

Only the direction in which they work with living beings is always determined by the need to preserve life, because life, unlike the phenomena on inanimate matter, is a temporary phenomenon that cannot be recalled.

The energies therefore always work in a purposeful way in the living being, but in contrast to inanimate natural bodies this requires a guiding principle, a manifestation whichmust be material in the living being because it cannot be measured. We call such a manifestation a bio-regulation system. It consists of the entirety of the information-processing processes. The processes combine useful information or process information with their carrier of biological-chemical substances.

The problem of life is thus synonymous with the problem of the bio-regulatory system that rules over the inanimate substances and energies required by life.

* * *

We have thus undoubtedly come much closer to the mystery of the temporal end; for to understand the temporal end, we must first of all know the relationship of the living to the inanimate.

But before we enter into the investigation of the temporal end itself, another investigation is necessary. It is not both the death of a plant or a mouse or a sparrow that forces such a burning participation on us, but our own temporal end, which we are inclined to assume from the outset to be different after all.

To establish this, our next question must be Is our life completely similar to that of other living beings? Is our being exhausted in what we have so far recognized as the principle of life? - This is what we shall deal with in the next section.

4. From awareness to information

Our investigation so far has led us to a clear separation of the inanimate from the living. The inanimate is the world of matter together with the energies that dominate it, which only work blindly and according to the law. The living is also bound to such matter, but it is also dominated by a bio-regulation system that works in a purposeful way.

If we now look at the world of life, three types appear to us in it: plant, animal and human.

If we first look for a common characteristic of all living beings, we find that all of them constantly (just like inanimate matter) interact with their environment, but that the stimuli from the environment acting on the living beings are answered by them in a way that is only for them and always appropriate: The cell with the protoplasm and the organelles contained therein (cf. p. 35), as the organized basis of all living beings, possesses, in contrast to all inanimate natural bodies, the ability of irritability and the ability to respond to stimuli in a purposeful way by any life activity.

The reaction of the cell to stimuli is unconsciously purposeful behaviour, and such action can be called innate behaviour. An important activity of the regulatory system is the organisation of such processes that respond appropriately to stimuli within the framework of innate behaviour.

The word "instinct", which has been used from time immemorial for the unconsciously purposeful activity of the innate behaviour of living beings, is only used cautiously at best in specialist literature today. Colloquially, however, the word is often used. When one thinks of its use, one thinks first and foremost of those wonderful animal actions which have always amazed people and often seem to be human actions. The most conspicuous behaviour is that of the sociable insects, bees and ants. But also with all other higher animals similar innate behavior shows up; the most wonderful behaviors are connected with the brood care, with the precaution for the future sex.

In all these actions, the expediency and certain success almost feign human consciousness, but this is completely impossible, because otherwise the same would in many cases surpass human consciousness, which is impossible to think of.

When the caterpillar of the peacock's eye at the entrance of the cocoon in which it wants to pupate makes a device like a weir that can only be opened outwards, this happens just as unconsciously as the removal of the funnel from the birch leaf by the funnel winder, which is done according to the laws of higher mathematics.

But the activity of the cabbage white butterfly is just as unconscious when it lays its eggs on a cabbage plant that it had never known itself, or when the female salamander climbs into the otherwise unfamiliar water every year to lay its eggs in it. It does not know that it was once equipped with gills for the water and that it underwent its first development in it.

These few examples from the almost infinite abundance of innate behaviour may suffice to clarify the term: Innate behaviour is an unconsciously purposeful activity that always aims at the preservation of life.

But we can add something else: The innate behaviour is hereditary. The children show it, without ever having learned it, in exactly the same way as the parents. Without this, there would indeed be a veritable confusion in the world of life. Yes, if the innate behaviour is to be a sure guide in the life of the creature, as it is indeed, then it does not have to depend on learning, but then it must be given to the living beings as an undeniable gift into life.

Even the first swimming attempts of a young duckling may be awkward, but it brings the ability to swim into the world, just as the chicken has the manner to scratch the ground with its feet. And if the caterpillar were first to be taught how to build the cocoon for its resting period, so that it would be a safe dwelling for it, it would not get beyond its first awkward beginnings and might become prey to its enemies. In short, innate behaviour must be a firmly and securely inherited ability if it is to fulfil its purpose of serving the preservation of life.

But now let us remember that after all, all processes of life basically happen in the same way: Inherited unconsciously purposefully and lawfully. In fact, if we think about it carefully, there is not the slightest reason to make a fundamental difference here, and so we can then regard all life processes as innate and, where it seems sensible, also call them innate behavior. The organization is done by innate processes or what the same means by innate behavior, the intake of food as well as the processing of food and the building of the body. And all these processes or behaviour are part of the bio-regulation system.

And we must also take the step of extending the term "innate behaviour" to plants. Or what should fundamentally prevent us from describing as innate behaviour the "searching" of the creeper for a support, which is also an unconsciously purposeful, inherited action, or the avoidance of dry or chemically harmful places in the soil on the part of the root, etc.?

And so it is that we must call this organizational and life-sustaining activity of the regulatory system in us humans innate behavior.

Of course, we do not deny that there will be a certain difference in the innate behaviour that will cause you to make a difference. It is perhaps justified to call those instinctive actions first mentioned and so respected since ancient times a more highly developed innate behaviour; but in principle it remains equivalent to the other fundamental processes in the cell as the outflow of an inherited, unconsciously purposeful activity.

So we find that we can speak of a regulatory system that organizes innate behavior in all plants and animals, and also in humans, in the sense described above.

Why animals decide about their behaviour

Only in the higher animals and humans, the stimulus responses are also expressed in another way, namely, insofar as these beings are equipped with a nerve-sense apparatus, in sensations which, responding to various external stimuli, raise life to a higher level. And where, on top of that, a trained central organ, a brain, is available, this innate behavior, a series of life expressions, shows through which it rises high above the organizing regulatory system, namely sensual ideas as well as so-called associations, i.e. connections of ideas. In the case of the latter, the new idea evoked by a sensory impression involuntarily triggers another idea previously associated with it; for example, in the case of a dog, the sight of the stick evokes the idea of the punishment previously associated with it; a certain smell evokes the idea of a certain person, etc.

In addition to both, i.e. to sensual conceptions and unconscious associations, there is also a sensual memory, for which the brain is, so to speak, the accumulator or information store. The associations in particular presuppose memory images. We can summarize all these abilities of the animal under the term "mind".

If we compare the behavior belonging to the mind with the information processing processes of our computers, we cannot avoid defining mind as a totality of information processing processes.

We often admire the actions of the higher animals, especially the dog. Even apart from the countless hunter's tales, which cannot claim to be credible, each of us has at one time or another come across a strikingly human-like action of an animal, so that one might be inclined to think of an intellect, of a human-like consciousness. We will have to talk about this right away. Here we must first of all note that even the most wonderful animal stories can be explained largely by what we have just presented as characteristic of the regulatory system of the highest animals: sensations due to existing sensory systems that react to stimuli such as light, sound waves, touch, pressure, muscle stretching, temperature, molecules of smell or taste. Furthermore probably also ideas, associations, memory images. Above all, we must not forget one thing: The sensory systems, and thus the sensations of animals, are often much finer than those of humans. Especially the dog, which so often surprises us with its almost deliberate actions, possesses olfactory sensors, of which we cannot even imagine in view of our much duller olfactory organ.

In the course of their lives, animals have to make a series of decisions: when to move and where to go, where to build a nest, what to eat, whether to fight or flee, with whom to form a community or with whom to mate. Without decisions their survival is left to chance. Making the wrong decisions reduces their biological fitness.

One of the most important decisions an animal has to make is the choice of habitat. The habitat must be a safe place for a nest, provide food in the environment and access to sexual partners. For example, seabirds choose cliffs or rocks in the sea to nest because these places offer protection from predators. Animals with a specific diet choose habitats where their food source is abundant. So animals use characteristics to select a suitable habitat.

For many species, the presence of conspecifics is an important characteristic for the suitability of a certain location as a habitat. The observation of conspecifics provides them with information on the quality of the habitat. Collared Flycatcher (Ficedula albicollis), for example, which arrive in their breeding areas in spring, regularly inspect the nests of their neighbours and other individuals.

Researchers suspect that Collared Flycatchers judge the quality of a habitat by how well the individuals in their neighbourhood are doing. Observation of the broods in the neighbourhood provides the birds with information as to whether there is sufficient food in the area or even a surplus of food. To test their hypothesis, the researchers enlarged the broods in some areas by taking young birds from nests in other areas and adding them to the test area. And indeed, the following year the Collared Flycatcher settled preferably in those places where the researchers had artificially enlarged the broods. The big broods were the characteristic for the birds for existing food-abundance.

Even if the choice of animals seems very simple in one example or another, the decisions cannot be dismissed as innate reactions to the strongest environmental stimulus. Undoubtedly, the animals first collect information before making a decision. This information must be processed. Information processing contains a process that leads to the recognition of observed features. Recognition means that a perceived object is determined to be consistent with a memory content. This memory content may be innate or acquired through previous experience. Recognition leads to an evaluation which exerts a more or less strong stimulus towards a decision.

Now it will be the case that a decision is not made immediately after the first best recognized feature. If no choice is made between different options, a species would have difficulty adapting to changing environmental conditions by making the best possible choice in the circumstances.

The observation of the animal world often shows a great adaptability to changing environmental conditions, so that we can rule out that animals do not make a real choice and are driven solely by chance.

Genuine choice means that at least two identical characteristics are recognized and evaluated in such a way that decisions are made on the basis of them. Two or more ratings lead to the same number of stimuli, which are compared with each other. The decision will be made for the strongest stimulus. One can consider such a decision as determined and explain it by "innate behaviour".

But even now the explanation of the decision is not as simple as it seems. Indeed, it will often be the case that two or more recognised characteristics lead to equally strong stimuli. There are then obviously no substantial differences in the recognized. For example, seabirds may not notice any significant differences between one rock or another that they inspect to decide on a nesting site. Or Collared Flycatcher may not notice any differences in the different broods in their environment.

Now there can no longer be a decision based on the strongest stimulus because there are no differences in stimulus strength. How the animal decides can no longer be predicted by any experiment. Once the animal decides for one habitat, another time for the other. The decision is not determined and therefore cannot be explained by "innate behaviour" alone. A further explanatory model is needed here.

We find such a further explanatory model when we grant animals a kind of consciousness. A definition of consciousness compatible with natural science can be found in the "Small Dictionary of NaturalPhilosophy[8]":

> *Consciousness is an information-processing process in which, in the event of new requirements or changed external circumstances, undetermined decisions are made between alternative courses of action that lead to targeted behaviour in order to satisfy needs. - A need is the tendency to pursue a goal.*

While this definition does not yet define higher self-consciousness, it does contain the minimum requirements that behavioural psychologists have established as criteria for the presence of consciousness.

If we compare the definition with what we know about the animals' decisions, we find that

1. It is an information-processing process that ultimately leads to a decision.

2. At least in the choice of habitat, there is a confrontation with new requirements.

3. If the stimuli are equally strong, undetermined decisions are made.

4. And finally, the decisions at least satisfy the supreme need, namely to preserve life. If you look more closely, you will see that a decision also satisfies subordinate needs (food, reproduction, security, etc.).

---

[8] Sedlacek: *Kleines Wörterbuch der Naturphilosophie: 1200 Begriffe, die man kennen sollte, kurz und prägnant;* Bod, Norderstedt (2016), Stichwort „Bewusstsein"

This means that the decisions that the animals have to make are conscious decisions because they meet the criteria for consciousness.

Behavioral psychologists have found out through experiments that one or the other species of animals (e.g. bonobos, elephants, ravens) even have a higher form of consciousness, namely self-consciousness, roughly comparable to human self-consciousness.

Regulation in humans

But the fact that we humans have a less sensitive sensory system than many animals has its good reason that we do not need it to the same degree as animals because of another higher function. And it is the same, by the way, with the innate behaviour, which is very much in the background in our case in view of that higher function.

Our regulatory system also reveals itself as an innate, organizing stimulus, namely in the construction and maintenance of our body; we must also assign sensory systems to it in the same way as the regulatory systems of animals. But is this the only manifestation of the human being? This is the important question that should concern us now above all.

The following consideration is fundamental to the question of what we consider to be the most important thing about ourselves: some people may want to exchange their external living conditions for others, a sick person may even want to exchange their body, but everyone wants to remain themselves, i.e. keep their "I".

This is a proof that everyone involuntarily sees his actual appearance in something other than his body, namely in what is called his "I", his "personality", his "self-confidence".

For this self-confidence one has rightly counted three functions since time immemorial: Thinking, wanting and feeling.

The first question that must now occupy us is whether, with this ego or self-consciousness, we have perhaps only reached a higher level of what we have recognized as a regulatory system in our previous investigations, or whether something fundamentally new is not to be seen in it after all.

The first possibility seems to be supported by the fact that humans have the same physical organ serving consciousness as the higher animals, the brain. The same shows a substantially higher education with it than with the highest ape, alone the difference to the ape-brain is nevertheless only a gradweiser, no fundamental one.

Now then one has indeed wanted to find those three spiritual functions of thinking, wanting and feeling in the animal, and has made every conceivable effort to find here a simple ladder from animal to human being. How about that?

We have seen earlier that we can speak of an "animal"mind and that the same is probably expressed in ideas and their associations. This activity was then also referred to as "thinking" of the animals and thus a bridge was built from the information-processing processes of the animals to human consciousness.

Sensual concepts, their connections and memory images are effects of the processes of biological regulation in the brain, and they are all the more perfect the more complicated the brain is, i.e. most perfect in humans. They alone do not yet deserve the title "thinking". Rather, they are only the material supplied by the brain and the regulatory system, which is only further processed in thinking by concept formation, abstraction and conclusions.

Here consciousness proves to be a sovereign, even creative force that dominates the material.

It is obvious that this activity is quite different from the purely linking information-processing processes of the animal, which we have described above as "mind". It is advisable to choose the word "reason" for the human mental activity, according to generally accepted expression.

As far as memory is concerned, we have already seen that we can find something similar in the direction of the highest animals; but the difference between animal and man is also obvious here, and so as not to blur it, it is better to distinguish sharply between "memory" and "recollection".

"Memory" is again a term with which one has made great misuse, whereby one has gone so far as to ascribe "memory" even to the substance as such. So: If the water only ever becomes solid at 0 degrees, this should be "memory". This is roundly said a nonsense, by which one blurs fixed differences between dead matter and life, as well as between matter and consciousness.

"Memory" is an information store together with the memory function of the nerve cells of the brain and nothing else. It provides for the automatic, involuntary repetition of ideas after new sensory stimulation from outside."Memory", on the other hand, is a process for the arbitrary reproduction of such memory images, even without external sensory stimulation. It presupposes a dominant activity.

As far as the will is concerned, this term is also used within the non-human sphere. Schopenhauer even extended it to the whole world (*"The world as will and imagination."*)

But if you do this - and it is doubtful whether it is advisable - then you must of necessity distinguish clearly between human and non-human will. In fact, both have a very different character. The animal will is completely exhausted in the satisfaction of its needs for the preservation of life; it is in complete dependence on matter and energy. Certainly, his phenomena are also a choice between two possibilities, a choice. The decision alone has as its final reason the need in matters of the preservation of life.

It can be said that it takes place in the direction of the physically strongest motive, and it is only in the case of two or more equally strong motives that undetermined decisions are made. If you put a piece of meat and a piece of bread in front of a dog, he will inevitably choose the meat, because that is where the stronger motive lies for him. The decision is different if, for example, he had to choose between water and bread, especially if the motives for satisfying hunger or thirst are about the same after long hours without eating or drinking. Through suitable experiments, one can find out how the strength of the motifs can be matched.

With the human will, on the other hand, it is something different, which is shown by the fact that man placed a completely new creation next to nature: culture. No animal has created culture and will ever create it; for with innate behaviour alone, that is, instincts, it is impossible. Shoots can never be creative.

Let us think of our example above: If you put meat and bread in front of a human being, he will often choose only the physically strongest motive, if he is not a vegetarian, and like the dog, he will choose the meat because it tastes better to him. He alone can choose differently based on personal and conscious convictions or cultural motives. In full awareness of his action, he can decide against the strongest physical motive - certainly determined causally and in this respect not "free" - therein lies the "freedom of the will", which does not mean an unrestrained or unconditional choice, but the ability to choose according to various motives, whereby the physical ones can completely take a back seat. Therein lies the dominant and higher-consciousness character of this kind of will.

And similarly it is then finally also with the third function of consciousness, feeling. One might be inclined to think of the emotional response to the stimulation of sensory cells, but that is not what we are talking about here. Feeling should rather denote a certain inner condition, the feeling of life of the being in question.

Surely the higher animals also possess such an attitude to life, which is expressed in lust and dislike. It is obvious that this refers to physical well-being. This makes it a pure function of the regulatory system, and as such this attitude to life will certainly be much more widespread in the world of life than we might be prepared to accept. Each of us knows this attitude to life and knows what is meant by it.

In contrast to the attitude towards life, feeling as a higher mental function has a different character, it is independent of physical things in itself, although it can of course lead to physical activities. Spiritual feeling is manifested in moral sentiments, especially in relations with the world of fellow men, and also in general ethical responsibility, whereby, closely related to the will, it shows its purely spiritual character in the mastery of drives and passions.

All these are characteristics by which the feeling, which differs from the attitude to life, rises far above the processes of biological regulation and thus also above animal life.

So we see that for those three higher mental functions of thinking, wanting and feeling, certain parallels can be discovered in the field of regulating biological processes, but that they can be easily separated from them, indeed, that they must be separated. Once again it should be emphasized: The processes of the bio-regulation system are always an unconscious phenomenon. The functions of the higher consciousness are a conscious and trend-setting phenomenon, which mainly influences the functions of the bio-regulatory system. It can only be confusing if one confuses these areas by using the same term for both, i.e. "thinking" also for the "mind" of the animal, the "will" also for the instinctive life of the animal and the "feeling" also for the pleasure and displeasure of the animal. One should limit those terms to the field of higher human consciousness, it would be much clearer.

In any case, these considerations have shown us that the human ego or personality is characterized by a number of features by which it rises above not only matter but also biological regulation in the sense we have outlined. She rules them both.

But what rules over the body and the bio-regulation system must be more than this, must be a special manifestation, must be independent of them in a certain sense.

This seems to be contradicted by a circumstance that is generally regarded as the most popular reason against the independence of consciousness: brain and consciousness are undoubtedly closely related. When the brain is asleep, the conscious mind seems to sleep when the brain is sick; the conscious mind also seems to be sick. It is concluded that with the temporal end of the brain (brain death) consciousness has to stop, just as seeing stops when the eye has died in the temporal end.

Apart from the fact that this view does not in the least do justice to the relationship we have just established between consciousness, matter and the regulatory system (soul), insofar as the matter is subordinated to the regulatory system and both are subordinated to the higher consciousness, we want to make ourselves clear that it is by no means the only possible, and therefore not the logically necessary view. The relationship of dependence could also be similar to that of the artist on his instrument. On a broken instrument, the greatest artist can only perform inadequately --- and without an instrument, he cannot even be heard.

So it may be that our consciousness is working with the brain in the same way as the artist works with the instrument. According to this, sleep and brain diseases cannot give us sufficient evidence against consciousness as an independent manifestation.

According to the previous general statements, it is now necessary to penetrate a little deeper into the essence of human consciousness.

Consciousness as a manifestation in its own right

Let us once again remember how our "thinking" takes place: The external things irritate our sense organs. The stimulus continues from the nerve to the brain and is "felt" here.

For all this, physical organs are necessary.

The "sensations thus obtained" generate ideas that can be linked together, from which we can then draw conclusions of general validity. This process is called induction. So we think with our normal consciousness -- let's call it day consciousness -- inductive.

In addition, the reverse path, the so-called deductive path, from the general to the particular, is also taken. However, we are still dependent on previous inductions, i.e. ultimately on sensory experiences and brain activity. This is what "logical thinking" consists of.

That then also our wanting and feeling during daytime consciousness is dependent on thinking is undoubtedly. And this is done in such a way that wanting and feeling are influenced by thinking, to a certain extent persuaded, so that our actions correspond to the experiences made by the sense organs and inductively generalized.

Such persuasion is called suggestion and, if it is done by ourselves, it is called autosuggestion. According to what we have just said, our daily consciousness is thus constantly under the autosuggestion of our brain's thinking, as we would like to call thinking that works with our sensual ideas.

This day-consciousness with its brain-thinking is the "intellect" and this is essentially the content of scientific psychology or psychology.

Some countries have gone beyond this and have included certain phenomena in their investigations, which are known as extrasensory perception (ASW).

Most scientists reject extrasensory perception. The reasons are mainly the lack of empirical evidence and the lack of a theory that could explain ASW. The main problem seems to be the lack of a suitable theory. Because without such a theory, phenomena that cannot be explained by classical theories cannot be correctly classified. Rather, scientists tend to dismiss such phenomena as fraud, humbug or even fantasy.

In this context I would like to remind you of the quantumphysical problem of the "spooky remote effect", for which there is still no explanation, although its existence has been proven with absolute certainty and in the meantime even technical applications of the phenomenon are being planned. What is this phenomenon about?

It is about the so-called quantum entanglement, which was denigrated by Albert Einstein as a "spooky long-distance effect" because he did not want to believe in its existence. This is a physical phenomenon in which, for example, light particles (photons) form a non-localconnection with each other and, in the case of a measurement over a huge distance, influence each other without time delay.

With the word "notlocal" the physicist somewhat bashfully disguises the fact that it is a connection outside of space and time, i.e. space-time. Within space-time it would mean "local". So, to say it once

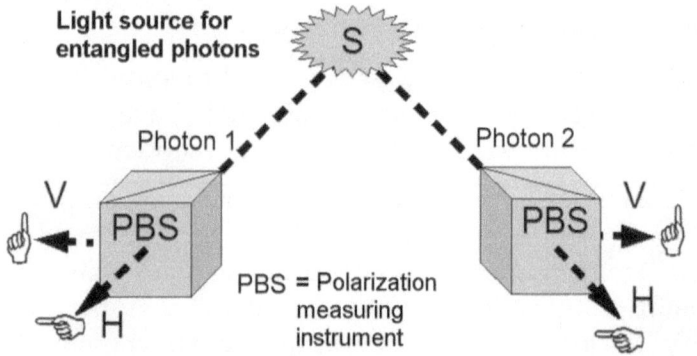

Fig. 4.1: If an H-polarization is measured at photon 1, a V-polarization is determined at photon 2. If, on the other hand, the V-polarization is measured for photon 1, the complementary polarization H is determined for photon 2. Neither polarization is fixed before the measurement. The effect at photon 2 occurs without time delay (instantaneously) faster than at the speed of light, even over huge distances. Graphic: Sedlacek

again quite clearly, "notlocal" means that it is a matter of a "beyond" of whatever kind, which exerts its influence on "this world". And then the physicist is silent, because he cannot and does not want to say anything about the "beyond". There's just no theory for it. Here once again the definition:

> *Quantum entanglement means the common state (entanglement) of a system of two or more particles, for example the common oscillation state (polarization) of two entangled light particles. Measurement results of the polarization of entangled particles are correlated, that is, not independent, even if the particles are very far apart.*

There are experiments where the particles are 17 km apart. If the polarization of particle A is measured, the correlated property, e.g. a polarization perpendicular to the polarization of A, is also found without delay (instantaneous) in particle B.

One could now assume that the properties and states of the entangled particles are fixed from the outset, so it is no wonder that the second measured of two entangled particles assumes the complementary state.

But this is not so. Before the measurement, entangled light particles have no polarization at all. Or to put it more precisely: they are in superposition of all possible states, they are in the superposition state, as the physicist puts it.

What does the term superposition state mean? To illustrate this, there is a famous thought experiment called "Schrödinger's cat". According to the rules of quantum mechanics, which are quite strange for non-physicists, in this thought experiment the cat is brought into a state in which it is simultaneously "alive" and "dead", and remains in this state until the experimental arrangement is examined. Only at the examination it is decided whether the cat is alive or dead.

Superimposition therefore means having all possible states at the same time. In the case of the cat, be dead and alive at the same time, as long as nothing is measured or examined. This is of course impossible in real life. It is just the expression that nobody knows anything about the cat's condition before the examination. As soon as you learn something about their condition, it is determined. Then the cat is only one of two possibilities. Either dead or alive.

Fig. 4.1: The quantum physicist Anton Zeilinger (2011) has experimentally confirmed the phenomenon of quantum entanglement ("spooky remote effect"). CC-BY-SA 2.5 Jaqueline Godany

This superposition also applies to entangled photon pairs before their measurement. Only when one of the photons is measured, the polarization is determined. If one wants, one can also say that at the moment of measurement the photon decides which polarization it assumes. After this decision the entangled second photon is also fixed. It assumes a complementary state. If the first photon chooses a horizontal polarization, then the complementary state of the second photon is a vertical polarization. The same applies to other decisions of the photon measured first. The behaviour of the second photon is caused by a "spooky remote effect" or non-local connection. The fact that we can name the phenomenon does not explain it. The physicists do not have an explanation.

Experiments carried out by the Austrian quantum physicist Anton Zeilinger and others in recent decades confirm the existence of the phenomenon.

If one wanted to describe the behaviour of lightparticles with a term of psychology, one would have to say: Particle B has perceived the measurement result of A extrasensory over very large distances and has communicated this to the physicist by delivering a similar (correlated) measurement result instantaneously, i.e. infinitely faster than light.

This is spooky, inexplicable, worthy of being called fraud or other unflattering things. No wonder Albert Einstein dismissed it as a "spooky long-distance effect".

But today it is the case that no serious physicist can doubt the existence of the phenomenon if he wants to continue to be taken seriously. The circumstances have almost reversed themselves compared to Einstein's time.

Back to the problem of extrasensory perception, which is rejected in the absence of a suitable theory. Here, the idea of a theory of ASW on the "spooky remote effect", i.e. quantum entanglement, as we want to say from now on, comes to mind.

The rejection of extrasensory perception by most psychologists may be understandable, however, if one considers the quite sober and critical character of our science and scholars and the fact that extrasensory phenomena are things in which fantasy and gullibility are all too easily flourishing. In addition, psychologists have usually never heard of quantum entanglement.

Another question is whether the behaviour of the scientific community is justified. The belief in extrasensory perception is so widespread that it must also be the task of science to examine its foundations and thus those phenomena with the greatest caution and criticism, but also completely without prejudice and without worrying about possible consequences for our views. If you do not do this, you are proceeding in an almost unscientific manner. Fearful distance from problems has often enough hindered science in its progress. And when one sees how little progress has been made for centuries in the scientific explanation of extrasensory perception, one cannot help but think that there are certain shortcomings here and that one should try out new approaches.

It is remarkable that, as already mentioned, in some countries, especially in England, France and Italy, acknowledged and important sober natural scientists have broken new ground and come up with striking results. It may suffice to recall men like the zoologist Wallace, the physicist Oliver Lodge, the physiologist Richet, the neurologist Lombroso and the astronomer Flammarion.

In order to see as clearly as circumstances permit in matters of consciousness and then further on the temporal end -- and this is the purpose of our endeavor -- it is essential that we go into these things; we want to do it objectively and soberly and without being influenced by left and right. Then we shall be saved from error. Unless I say otherwise, I will only report those things that are based on certain facts after careful examination.

If we consider the matter at hand, we must say the following, even after our previous research: There is a probability that consciousness is a manifestation that is independent of the brain in a certain sense. Should this really be the case then we will not find the truth if we always try to approach consciousness from the brain only, as psychology and other science mostly still do.

If this independence of consciousness really exists - and science must consider this possibility - then we must rather, when examining consciousness, first and foremost take into account such situations and conditions in which the brain is more or less switched off, so that any independent consciousness of man can emerge more unmixed and pure. This is so undoubtedly right and natural that there is really no need to say a word about it, and that one only has to wonder if psychologists or other scientists do not take this promising path, or even reject it.

Everybody knows about shutdowns of normal brain activity. First of all, natural sleep belongs there.

During sleep the body rests, and with it the brain and senses, whether the latter is paralyzed by "fatigue substances" or otherwise. We are then no longer able to think inductively. Our daily consciousness is switched off - Is the consciousness also at rest, without activity?

In any case, it is not always completely without activity in sleep. The dreams prove it. Of course, these are mostly based on peripheral irritations of the brain and on ideas of the sleeper in the waking state, which are then repeated in a confused and disorderly manner and bear the mark of the impossible on the forehead. It is known that this is due to the dormant state of the brain and that other physical conditions also influence these dreams. They represent the ordinary kind of dream, and I would like to call them "confused dreams" because of their confused character.

Now there are other types of dreams, especially one that I would like to call "clear dreams", in contrast to the "clear dreams" just mentioned, because they neither formally have anything confusing about them, nor do they deal with impossible things, situations and images. Such clear dreams are the subject of very general experience for all those who are accustomed to working intellectually productively. Be it in deep sleep or half-sleep, they bring immediate, i.e. without the usual process of inductive thinking, insights, problems, dispositions, yes, in the case of artists, poets, composers sometimes a detailed presentation of performed works or parts of such.

We will still have to talk about the fact that such enlightenments, which are not gained through the usual work of thinking, but are, as it were, given as a gift - they are called intuitions - also very well occur in the awake state, in day-consciousness; but the important thing for us is that they occur nevertheless in sleep. Since they take place with considerably reduced brain activity, they are a valuable indication of the possible independence of consciousness, of its effectiveness beyond or above the activity of the senses.

Handel had a very remarkable lucid dream at the end of the work on his "Messiah". He was unable to perform the final chorus, the great and mighty "Hallelujah!" and therefore went to bed one evening tired and discontented. During the night he clearly dreamed the solution of the problem he had in mind, which he had not found during the day, and when he awoke in the morning he immediately noted the magnificent composition of his lucid dream, which has since then had the effect of a revelation from higher spheres on so many people.

He also completed a violin sonata by Tartini.

After he had tried in vain, he fell asleep and dreamt that the devil had appeared to him and offered to complete the sonata if he gave him his soul in return.

Tartini responded, the devil took hold of the violin and played the sonata to the end with the most delightful magic. When the master awoke, he immediately wrote down the melody heard in the dream. -[9] What happened to his soul is not recorded.

Condillac says that he often solved problems that were bothering him before he went to sleep while he slept.[10] - Similar cases are often reported and everyone will have more or less experienced them themselves.

The philosopher Du Prel mentions a very remarkable case of lucid dream, which is complicated by the fact that it is at the same time about the extrasensory perception of distant things:

> "A scholar in Dijon fell asleep after vain attempts to understand the sentence of a Greek poet. In a dream he was transferred to Stockholm to the residence of Queen Christine and was placed in front of a compartment in the library, where his eye fell on a small volume which he opened and in which he found about ten to twelve verses and thus the solution to the difficulty. He woke up joyfully and noted down what he read, then wrote to his friend Chamet, the envoy in Stockholm, asking him to ask the philosopher Descartes there about the establishment of the library. He enclosed the copy of the verses read in the dream and asked to see if these verses could be found in a particular volume of a particular subject. Descartes thought everything was right and said that "you can't give more precise proofs if you've been visiting the library for 20 years.

Somewhat different from the clear dreams is the so-called "true dreams", which are dreams that present

---

[9] Flammarion, *Rätsel des Seelenlebens*- Stuttgart, J. Hoffmann 1909, p. 291.
[10] U.O.S. p. 36

scenes that happen just now, often in the far distance. The extrasensory perception of a simultaneous event is called clairvoyance. We are in a different position in relation to such dreams of truth, because they are much less the subject of general experience than clear dreams, but are reported to us by individual personalities whose credibility and critical aptitude is always important.

It is therefore quite clear that one must always confront dreams of truth with strong criticism. On the other hand, it is also not acceptable to doubt the cases that occur without any reason. And at last it is not that it is something completely unbelievable.

There is one more thing in particular to consider. The number of such dreams of truth may be relatively small, but isolated cases alone show a great deal. If the credibility of the persons in question is beyond any doubt, then, if one does not want to accept these dreams as proof of the nature of consciousness, all that remains is the assumption of a coincidence.

However, if it is a dream that reproduces an event in all its details, then the probability of the dream and the actual event coinciding is so unlikely according to the rules of probability theory that it cannot be a coincidence. If there is no other explanation, should the integrity of the rapporteur be questioned? I don't think that would be a solution worthy of science.

Of the many dreams of truth reported by credible people, there is only one case that can be considered typical and which is guaranteed by the unimpeachable credibility of the rapporteur.

Stutzer reports the following in his magnificent work *"In Germany and Brazil"*[11], which was published in 1913 and had four editions in one year:

> "My mother, the healthiest woman and nothing less than nervous, wakes up one evening with a terrible scream, can't calm down at all and then says to my father: "I just saw our Otto fall off a horse off a rock. (My brother had been in Brazil for four years at the time.) Father was very sober about such things, said it was a silly dream, but looked at the clock silently; it was 11 o'clock in the evening. He wrote the incident in his calendar with the date, 21 May 1859. After three months a letter arrived from my brother, in which he wrote, among other things, that he was very well, but that he had had to overcome a bad case. He had ridden up the Sierra (the high coastal mountains) with several people on May 21 -- to buy horses on the Curytibanos plateau. The very narrow mule track, which was only wide enough for one animal, led hard past steeply falling rocks. They would have wanted to reach the height before total darkness. It was already 7:00. There his riding donkey had stumbled over a stone, he had flown out of the saddle, down the rocky slope, and would have been infallibly lost, had not a tree trunk stopped him. - The time difference between southern Brazil and Germany is four hours; 11 o'clock in the evening in Semmenstedt is 7 o'clock in the evening in Blumenau.

Even if we do not want to acknowledge the "true dreams", the result of our investigation of natural sleep remains unaffected. Sleep can be understood as one likes, according to the "fatigue theory" it is regulated or produced by biochemical influences of messenger substances and hormones, it is always an inhibition or elimination of parts of the brain and its actual activity, which is already sufficient from the cessation of daytime consciousness. And now we see that our consciousness is nevertheless not in inactivity, but that it often works creatively, and can sometimes even develop a more significant activity than in day consciousness.

---

[11] Brunswick. H. Wollermann, 4th ed. 1914, p. 72.

There is one unfortunate thing about these facts, that they cannot be determined arbitrarily by experiment, as is the case with the exact sciences; for normal sleep is not always interrupted by such dreams, but one is in the very favourable position of being able, at least within certain limits, to apply the experiment in this field.

For there is an artificial sleep, the so-called hypnosis, which is created by looking at a shiny object, by touching it, or even by simple command, for some people it is easy, for others it is more difficult, for some it is not at all. People behave very differently in this, as in all other things. That is why one should not be surprised that people are so different in terms of their capacity to absorb, etc., the extrasensory things discussed here. Nor can all people paint, compose or make inventions.

During such an artificial sleep the hypnotized person is completely under the suggestion of the hypnotist, and the deeper the hypnosis, that is, the sleep, is, i.e. the more completely the different layers of the brain are switched off, the more remarkable are the phenomena that occur.

The hypnotized person then follows the hypnotist's intuitions completely: if he is told he is an "emperor," he gives himself the dignity of one; if he is persuaded that he is a "beggar," he behaves like one. He jumps around croaking like a frog when he is told he is one; he is drunk by water that is given to him as "brandy". Even skin reactions occur; because under a sheet of stamp paper a draught bubble develops when the hypnotist has put it on as a "Spanish fly".

Above all the hypnotized person shows three more strange qualities: He can often read the contents of the consciousness of others, he has an astonishing memory, so that he e.g. executes orders to the minute, even in the waking state, and he is very receptive to thought transmissions. We'll talk about that later.

As far as the physical state during hypnosis is concerned, i.e. the state of the brain, the same is considered to be "functional latency", certain parts of the brain are "inhibited", i.e. switched off, just like during sleep.

Here we can let the surgeon Carl Ludwig Schleich have his say.[12] Schleich has developed a method of local anaesthesia and in this context has investigated the inhibition or dimming of the individual layers of consciousness in self-experiments. In addition, he has found that in hypnosis the individual layers of consciousness are gradually inhibited in the same way as in anaesthesia.

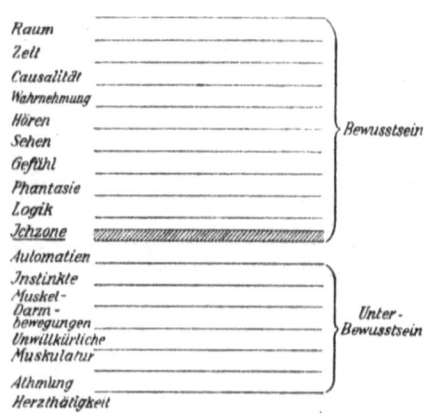

Fig. 4.2: Diagram of the functional layers of the cerebral cortex up to the deepest layers of the brain as revealed by anaesthesia.

---

[12] C. L. Schleich and Klaus-Dieter Sedlacek (eds.): *Bewusstsein und Unsterblichkeit*; Norderstedt (2015).

Now it is interesting to observe, when one is anaesthetised, that the first layers that are inhibited are those that transmit the concepts of space and time (Fig. 4.2). The orientation of time is first attacked, the concept of space blurs, then comes causality - cause and effect. Here we already have a strange result, which contrasts sharply with the philosophy of Kant. Kant built the critique of pure reason on the so-called a priori concept that we possess, with which we are, as it were, born. A misunderstanding can arise if one assumes that these a priori concepts are too deeply rooted in the human spirit, as it were in our mental substance, and now anaesthesia proves that space and time are among the very young shoots of the brain tree.

By the way, I would like to point out that this fact gets a strange reflection in Einstein's theory of relativity, which also includes an attack against the previous concept of space and time. But this leads us too far away from our core point. So be content with the fact that space and time and causality are the first things to go out with every anaesthetised person.

Then comes the imagination, the logic, the perception, the hearing, the seeing, the sense of touch. And now, when my imagination has made the biggest leaps due to the stimulus of the anaesthetic, when the logical relationships have ceased and my general sense of perception has gone out, the reduction of consciousness continues layer by layer. In the attack against fantasy, the ability to dream is first very vividly stimulated before it is turned off. In the attack against logic, one believes in a dream to have solved things that otherwise would not be solvable consciously. I myself once believed to discover a great mathematical system in the dream fantasy, which later dissolved into vain haze. Then the feeling of perception is perverted, hearing is transformed into hallucinations, until all sensory impressions are also silent, until, as it were, earlids are lowered in front of our hearing ability, as nature has placed them before our eyes. Then comes the sense of touch. The pain does not stop, it is muffled, while horrible ideas are conjured up by the imagination, the fantasy shoots around like ants and lizards. And only now, in about the tenth place, the ego disappears into the sea of oblivion and absorption. What is underneath is all activity of the subconscious. ...

Another condition that stands between the pathological and the normal is hypnosis. Here, any exposure to substances is eliminated. One only needs to stroke a dispositioned person over the forehead - the consequences are variable, but there is indeed objective evidence for the reality of hypnosis - or to look sharply at him, to address a friendly or harsh word to him, the brain zones up to the ego are suddenly dimmed. Orientation for time, space and causality fall away in hypnosis as well as in anaesthesia, then the other attenuations do not follow. In the depth effect of the brain inhibition, hypnosis only goes right up to the ego; but the ego as a zone remains (Fig. 4.3).

Fig. 4.3: Cerebral dimming in hypnosis to the ego zone.

This means that the ego lies bare like a muscle under the knife cut. The spiritual epidermis is separated, underneath the deeper zones are exposed, the zone of the ego is exposed. It is now, as it were, a piano on which a foreign will plays. The foreign will is directly accessible to the ego by eliminating the orientation, so that even exits that are made are automatically carried out by the exposed ego during hypnosis without further ado, as if the own consciousness of all dimmed zones let the ego act.

Here is the solution and the key to the core of hypnosis. After the onset of hypnosis someone other than the ego can play the piano on our apparatuses. This is all the more remarkable as there is no question of biochemical action by simple brushstrokes, words or glances.

Most important are the phenomena of somnambulism[13] (sleepwalking), which is unfortunately almost completely neglected by psychologists. One simply regards him as pathological and thus believes to have dismissed him as useless for normal psychology. This alone makes one forget a very significant fact. The somnambulists show extremely wonderful phenomena of increases in the activity of consciousness, which cannot be explained by the normally working brain. How then can they be explained by a sick brain?

---

[13] **Somnambulism,** or **sleepwalking**, is a phenomenon where the sleeper walks around without waking up and sometimes performs activities.

Somnambulism is a deep sleep-like state that occurs either by itself or through touch and hypnosis.

Day-consciousness is diminished in the somnambular state, but very peculiar abilities appear which the person does not possess in the waking state: Speaking in foreign languages, poetry, painting etc.

Above all the somnambulists have a most wonderful gift, which is called extrasensory perception. For example, the Munich philosopher Huber reported in his lectures on psychology[14] that a somnambulist in Tyrol described his room in Munich in great detail. However, she indicated the position of the furniture quite differently from what the philosopher knew, and stubbornly persisted in doing so in spite of all the counter-conceptions. - On Huber's return to Munich it became apparent that his wife had arranged for this change in the position of the furniture during his absence. At first the somnambulist also stated the number of windows in the room to be two; but when Huber said that this was incorrect, she replied: "In fact, it has only one, the other is a fake window.

A neurologist reported a painter who painted best when he was asleep. A journalist had taken on an essay for a magazine that he did not want to succeed. Finally he threw the manuscript away and wrote to the editors that he did not feel up to the task. To his astonishment, he simultaneously received a letter from the editors confirming receipt of the promised essay, noting that it had been very well done. When he was shown the handwritten manuscript at the editorial office, he actually recognized his own writing: he had written it in a dreamlike state.

---

[14] Cf. M. Perty: *„Ohne die mystischen Tatsachen keine erschöpfende Psychologie"*, Leipzig, E. F. Winter-, 1883, p. 27.

A chess player solved the most difficult chess problems in sleepwalking and a preacher worked out his best sermons in this state.

Above all, two very peculiar abilities still occur in the somnambulists as in hypnosis: Transmission of information between living beings without the involvement of the known senses (telepathy) and amazing memory capacity. Given the great importance of these things, we must briefly touch on them.

Attempts of the Russian researcher Dr. Kotik, fraud was probably excluded.[15] He did the first series of experiments with a fourteen-year-old girl and her father. When the girl touched her father, she very quickly guessed at objects that were familiar to her, but she was slower and spelled out when the object and its name were unknown to her; even when Kotik touched her, she was successful, but not with other people.

A second series of experiments was made by Kotik with a very reliable somnambulistic young girl. It is remarkable that the medium's answers always depended on what the participants in the session thought and felt. The medium was finally able to read even sealed letters, and the reproduction sometimes even contained more details than the letter, whereupon the letter writer explained that these did indeed correspond to what he had experienced, i.e. what he had kept in his memory.

Professor Schottelius, a physician in Freiburg, probably made similar observations on the transmission of information without the participation of the senses.

---

[15] Dennert:*Gibt es ein Leben nach dem Tode*; Godesberg (1915), p. 44.

Another mental characteristic that one has the opportunity to observe during these phenomena is a quite amazing memory capacity. But the same also occurs otherwise in pathological processes, often in an almost perfect education. Two examples of this.

A farmer's wife lay in violent feverish delirium and spoke in foreign languages so that her relatives thought she was possessed by the devil. But the doctor realized that she spoke Greek and Hebrew. Later it turned out that as a girl the woman had been in the service of an old priest who had the habit of walking up and down the long corridor of his house, reading Greek and Hebrew books. The doctor sought out the latter and actually found in them the passages which the woman, who had never learned these languages herself, had repeated in her deliriums.

A very similar case is reported by a Danish neurologist Dr. Lehmann[16] about a Danish girl who, in a state of dissociative disorder, gave long Swedish sermons in not quite correct pronunciation. She had never learned Swedish, but 10 years earlier she had been a domestic servant of a Danish parish priest who was studying Swedish and read aloud in a Swedish sermon book with not quite correct pronunciation.

We have treated in the previous: Clear dreams, dreams of truth, hypnosis, somnambulism, delirium. With all these phenomena, one circumstance is of resounding importance, because it can be observed again and again. These phenomena occur all the more surely, the more thoroughly the activity and involvement of the known sensory channels is switched off, as is the case in sleep, hypnosis, feverish deli-

---
[16] See Stutzer, *Geheimnisse des Seelenlebens.* Brunswick, Wollersmann, 1915, p. 120.

rium, somnambulism. Those phenomena are most striking in the deepest somnambular sleep.

So this must be a special kind of recognition. Normal recognition is achieved through brain activity based on the ideas gained through sensory perception, as we have already discussed. With the special type of recognition, sensory perception is excluded and the upper functional layers of the brain (Fig. 4.2) are more or less faded out. Nevertheless, the consciousness then continues to work in such a way that it often surpasses the activity of the day consciousness or a particular intuition comes to light.

This includes artistic and often scientific work. This is so important because this is where it occurs in a normal state. There is hardly a great artist and poet who does not acknowledge this. Thus a sober man like Goethe calls his thoughts "pure children of God" and "unexpected gifts from above".

We find a highly significant passage in a letter from Mozart[17], which I would like to repeat here for the sake of importance:

> "When I am right for myself and in good spirits, for example when travelling in a car or after a good meal when going for a walk, at night when I can't sleep, my thoughts come to me streamily and best. How and how, I don't know, is something I can't help. Those I like now, I keep in my head and probably hum it to myself, as others have at least told me. If I hold on to this, I soon learn one by one what such a chunk is used for, to make a pie out of it, according to counterpoint, according to the sound of the various instruments, etc. That now heats up my soul, for if I am not disturbed; then it becomes bigger and bigger, and I spread it out further and further and brighter, and the thing really becomes almost finished in my head, even if it is long, so that I overlook it afterwards with one look, as it were, like a beautiful song or a pretty person in my mind,

---

[17] Jahn, Mozart 3rd vol., p. 423.

> and also do not hear it at all one after the other, as it must come afterwards, in my imagination, but everything together. Now that's a treat! Finding and doing all this is only like a strong dream inside me; but not hearing it, all together, is the best thing".

Note the expression "like a beautifully strong dream", which places this concept of artistic genius in parallel with the phenomena of hypnosis and somnambulistic sleep.

In connection with this is the highly striking ability of children with island talents[18] in the musical and mathematical field. Mozart in particular had been such a musical prodigy since his early youth. Before he learned to write, he composed a very difficult piece and at the age of six he played the concert pieces of the greatest masters.

One of the most wonderful calculatinggeniuses was Zerah Colburn, who was born in Cabut (North America) on September 1, 1804. At the age of six he was able to calculate powers and roots so quickly that the co-writers could not follow, for example the cubic root of a nine-digit number. A gentleman asked him to name the factors of the number 247483, there are only two, 941 and 263, and he named them immediately. Before the question could be written down, he named the minutes and seconds of 48 years, an eight-digit and ten-digit number respectively. - How the number got in his head, the boy did not know how to say. On paper he could not carry out the simplest calculations. - For some things that he easily recognized immediately, such as the character of

---

[18] **Island Giftedness**, also known as **Savant Syndrome,** is the phenomenon whereby people with a more or less strong specific cognitive peculiarity are able to achieve extraordinary performance in a small sub-area ("islands"). At present, there are about 100 people worldwide who are known for their extraordinary and highly amazing achievements.

large numbers as prime numbers (i.e. indivisible numbers, the example given is 36083), no rules are known at all. So it cannot be a matter of memory.

When the boy later received proper arithmetic lessons, his wonderful ability diminished in proportion to the effort put into it.

It is even more striking that people with considerable intellectual disabilities are often quite talented in music and arithmetic.

* * *

Now one could make the objection with everything that has been said that it is not a matter here of phenomena that apply to all human beings, that is to say, phenomena that characterize human consciousness in general, but rather of individual human beings, with whom we can observe them.

But this objection is only apparently correct, and it is precisely the last thing said, artistic and scientific intuition, that leads us to a general characteristic of man. In fact, every human being has a more or less lively intuition in the form of the imagination, which is free and creative with the material provided by the imagination, and there is also something of the inventive spirit, which is genuinely intuitive, in every human being.

Here lie the deepest reasons why an abysmal gap yawns between the highest animal, which has only a regulatory system and a small amount of "mind", and the simplest human being, who has an intuitive consciousness next to it.

The human consciousness is - we have emphasized it again and again - a creative force. It masters the material that provided it with substance and regulatory system, and creates something new from it that is not in the pathways that heredity and senses have prescribed, as is the case even with the highest and wisest animal. Especially fantasy and inventiveness, which is the privilege of all people, prove this in the clearest way.

* * *

Let us learn the lessons from all these phenomena. Our research shows that we have another consciousness activity besides what we called "day consciousness" or just "consciousness". Modern psychology has not completely passed over her, it is what she calls "subconsciousness".

How are the two different?

1. the day consciousness is certainly dependent on the sensory perception, whereas the subconsciousness is only indirectly dependent. The clearer it becomes, the more its independence from the senses becomes apparent.

2. the daily consciousness is dependent on the waking state, but the subconscious does not. Because as soon as the waking state is switched off in any way, the day consciousness disappears while the subconsciousness is still active, or only then becomes particularly apparent.

3. the daily consciousness works and thinks laboriously inductively, drawing general conclusions from the individual sensory experiences. The subconscious, on the other hand, works effortlessly deductively by deducing something special from general sentences.

4. day consciousness is little or not at all influenced by foreign suggestion, it is constantly under the influence of the sensory experiences or the induction from them. On the other hand, the subconscious is cer-

tainly subject to suggestion, so that it then concludes its deductive conclusions even against its own experience and reason, but always logically.

5. the day consciousness has a limited memory and recollection capacity, whereas the subconsciousness has an almost perfect one.

6) In the documented cases, the subconscious probably possesses the ability of extrasensory perception and can transfer information between living beings without the involvement of the known sensory channels. This ability is completely lacking in the day consciousness.

7. the subconscious mind has the gift of intuition, which is hardly shown during the activity of the day consciousness.

In addition to the seven points, there is an eighth:

8. the day consciousness is the area of the calculating mind, the subconsciousness is the area of feelings.

If we wanted to separate the two types of consciousness sharply from each other, then the day consciousness would have the sensual experience and induction alone, while the subconscious would have intuition and extrasensory perception alone. However, such a sharp separation is not possible because they both have, albeit to different degrees, common abilities, such as deduction and memory.

A closer examination shows that day-consciousness increases with the strengthening of the body and brain and decreases with its weakening and aging, that on the other hand the subconsciousness often increases with the dwindling of physical strength and with weakening or obstruction of conscious brain activity, yes, certain phenomena perhaps indicate that it is strongest in the hour of the temporal end.

From this we may further conclude that the daily consciousness is primarily bound to the brain and its ability to work, but that on the other hand the subconsciousness is possibly independent of the brain and thus of the body and substance to a certain extent.

* * *

Now to the question of the relationship between the two types of consciousness and the principle of life, i.e. the regulatory system:

We have already seen that we cannot separate the two types of consciousness sharply from each other, because they have common abilities in addition to separate ones. The subconscious is the field of creative thinking, of feelings in the higher sense, it is what we see as the personal expression of the human being.

We will have to talk about the meaning of everything said here in the third part of this Scripture. Here is just a brief statement:

The human consciousness is connected to the body, and thus helps to develop its personality. Above all, the world of matter has a high intellectual-historical significance: it provides the necessary environment for the development of personality.

Since consciousness and matter are very different manifestations, it is necessary that they are linked by a binding agent; the binding agent is the processes of the bio-regulatory system, as we already learned in the first section. Since consciousness itself is an information-processing process and information, as mentioned in the first chapter, requires a material or energetic carrier in our space-time[19], the processes

---

[19] In the theory of relativity, space and time are combined to form a uniform

are of special importance as the connecting element between material and information.

Processes consist on the one hand of control information and on the other hand of substance, and both together cause physical interactions. An interaction is known to be that which moves, changes, transforms or causes another material reaction.

We saw that a main activity of the regulatory system is the organization of the body. With increasing development, it has progressed so far that it has been able to form the human body and in it the most wonderful and complicated organ there is, namely the human brain with the sense organs.

A tool is created in the brain, which the consciousness can use with the help of the bio-regulation system, in order to communicate with the outside world of the substance and develop on it.

It is not necessary to think about two separate manifestations of consciousness in the daytime and subconsciousness, for example, and thereby unnecessarily complicate the theory. The day-consciousness or the intellect with its induction is thus merely an exercise of consciousness, in so far as it uses its physical instrument, the brain. With the inhibition of the uppermost layers of consciousness, especially the dimming of the space, time and causality zone, the subconsciousness emerges in its particularity.

Consciousness is therefore a uniform manifestation, which only appears as day consciousness in its material connection with the brain.

What about the relationship between consciousness and the regulatory system?

Also the regulatory system, in contrast to the substance, is amaterial principle as we know it from in-

---

four-dimensional structure with special properties under the name of **space-time**.

formation processing systems. The transition from the subconscious to the regulatory system is fluent. A clear separation of the areas is not possible here either. One must even assume that many functions of the regulatory system can be regarded as information-processing functions of the subconsciousness without great problems.

The activity of the regulatory system is above all a purposeful one and proceeds in innate ways. - However, we do not find in it the characteristic features of the higher consciousness including the subconscious, namely: creative intuition, inventiveness, imagination, culture and extrasensory perception, and above all ethical will. These characteristics mark something fundamentally new in human consciousness besides the regulatory system. An unbridgeable gap yawns here.

Can ASW be explained scientifically?

The possible independence of consciousness from matter or matter is an important question to discuss before finding an answer to the question of life after life.

We have found indications of an independent consciousness in various reports of extrasensory perception.

Let us assume that extrasensory perception exists just as there is nolocal remote effect in quantum entanglement. Then the question arises how the consciousness, in this case probably the subconsciousness, got hold of the perceived information. The sensory organs for hearing, seeing, smelling, feeling or tasting may not have provided the information if the perceived object is far away, sometimes even thousands of kilometres. Sensory organs receive information from processes only over short distances.

Could some part of the consciousness have traveled to the place of perception instead, and at tremendous speed? There is no plausible explanation for a travel hypothesis in the fund of already existing theories of our present physics. Therefore I would consider journeys of consciousness to be extremely unlikely.

But there are phenomena in quantum physics which may possibly bring about extrasensory perception in a similar way as the processes in quantumphysical experiments. In the experiments on quantum entanglement, two entangled light particles communicate over any distance via a non-local connection, i.e. infinitely faster than the speed of light. Here, one last time, one is reminded of Einstein's term "spooky long-distance effect".

The prerequisite for a non-local connection between two particles is a previous interaction between them. Because only through an interaction the state of entanglement is reached.

In the life of humans, there are always interactions between relatives and acquaintances. One touch, for example when shaking hands, is enough. Forgive me for describing human relationships in physical terms, but it is only in this way that it is possible to build up an understanding of what happens physically.

Once there has been an interaction between two people, each person possesses entangled particles, atoms or molecules whose entangled twin can be found in the other person. Let us further assume that the entangled particles pass on their entangled state within the body, for example through interactions with blood cells.

The assumption that in two people each has one of two entangled particles is not a spinning but biophysics, as experiments to explain the high efficiency of photosynthesis in plants show.[20] Nature obviously uses quantum entanglement for a large number of biochemical processes. Interestingly, the plants seem to succeed in maintaining the interlocking states for a sufficiently long time and in constantly regenerating them. Why should this not be the case with humans?

Once two people have intertwined blood cells, it is no longer a long way to connect with the information processing processes of the subconscious. Because the brain, in which information processing is carried out, also needs the oxygen that the blood brings to it. And to the oxygen the entanglement has been passed on before.

If one of the two persons mentioned then experiences an extraordinary experience, this can be communicated to the subconscious of the second person without time delay by means of quantum entanglement. The knowledge about the experience of the other person is not yet decoded and in full clarity, but the subconscious connects it with other information stored in the memory, which works like a decoding. Finally, the result rises in the subconscious, either into dream consciousness or as a kind of uneasiness into day consciousness.

If the connections made by the subconscious were good and plausible, the communicated experience

---
[20] cf. Wrobel and Sedlacek: *Leben aus Quantenstaub*, Norderstedt (2014), p. 69 ff.

can even appear in the day consciousness with frightening clarity instead of as a dull feeling. However, whether what has risen so high in consciousness belongs to an event that actually took place can only be confirmed afterwards with the help of classical communication and information.

We now have a theory for extrasensory perception, but still no explanation for what a non-local connection means. As already mentioned, mainstreamphysics is shirking an explanation. It copes well with its current status. For those of us who are searching here for a scientific answer to the question of what happens at the end of time, the attitude of mainstreamphysics is unbearable. And so we have no other choice than to help mainstreamphysics a little by discussing and finding answers ourselves.

Is there a real afterlife?

In order to save ourselves from metaphysical adventures, we would do well to keep in mind an important result of epistemology, which can only be briefly mentioned here without further explanation.

It is the insight that all concepts of science are only in relation to the realities they denote, and that all our vivid ideas, in which we think the concepts of natural things and processes, are only to be understood as illustrations, in which we imagine the conditions of reality, but that they are by no means copies, through which we take things into our consciousness.

Thus it is particularly important to note that our vivid ideas of spatial relationships are something quite different from the objective spatial order of physical objects. The former are assigned to the latter, there is no other type of correspondence.

The movements of particles, e.g. of electrons, take place in physical space, and this is something quite different from the things we directly experience in our perceptions as "spatiality" or "expansion".

The descriptive space is a completely different one for the data of the sense of touch, the sense of sight, the sense of hearing, etc., so there are several descriptive spaces; the physical space, on the other hand, is only one, it is an objective order and, unlike those, is completely unimaginative.

Imagine a person who has been blind from birth and who has to find his way back with his sense of touch and hearing. This person will experience space as something quite different from the person who has full sight. Bats experience space by means of the ultrasonic signals they emit, and the proportions and views of space that insects perceive through compound eyes would amaze us humans. But no matter how space is subjectively experienced, none of these vivid, sensual spaces correspond to physical space.

Its properties can only be described by mathematical terms, can also be illustrated by sensual ideas, but are not something imaginable in themselves.

Physics had felt compelled to introduce as the last elements of reality completely non-descriptive quantities (e.g. electric field strengths, quantum fields, etc.); the epistemologically oriented person must now take the step of recognizing the physical image of nature in terms of its spatial and temporal form as a thoroughly non-descriptive entity that is constructed from abstract concepts and of which any descriptive properties cannot be meaningfully expressed.

We should always keep this in mind when I now try to find an explanation for the non-local connection of entangled particles as understandably and clearly as possible.

Every scientific consideration of reality - like, incidentally, all practical behaviour in the world - is based on the assumption that all events in nature take place according to law.

And there are no effects in the temporal distance. All dependencies can be traced back to those between processes that are directly adjacent in time.

In classical physics, we can add another statement to this: "What happens at any point in nature in a small period of time depends only on what happens in the nearest spatial and temporal neighborhood of that point.

But this last statement of classical physics obviously does not apply to the behaviour of entangled particles, because for such particles there is a spatially distant effect, or as one can also say, a nonlocalconnection.

Non-local connection means that a mediation of the interaction between the entangled particles takes place in non-local space.In this non-localspace, distances (= spatial extension) play no role, as experiments have shown. So nothing can exist in non-local space that has any spatial extension. Even element-

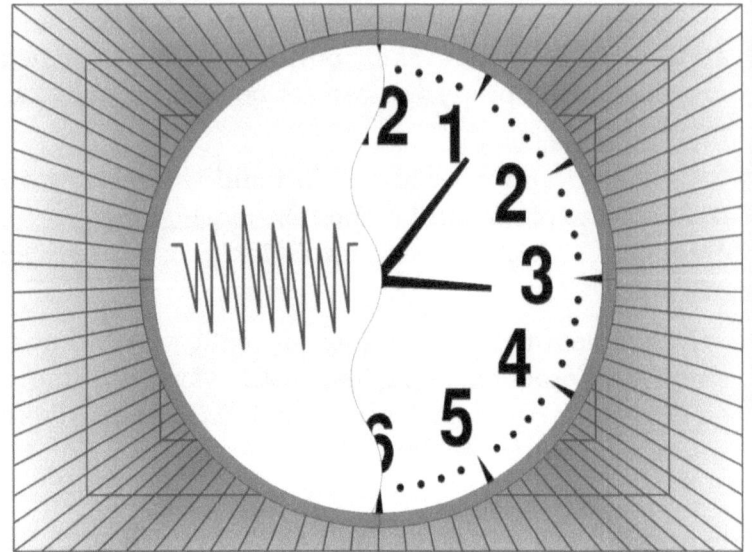

Fig. 4.4: Symbolic representation of the physical spaces this side (= space-time) and beyond (= metric-free vacuum). In this world time and spatial expansion exist. In the hereafter there is neither time nor spatial expansion, only information and processes. Graphic: Sedlacek

ary particles like electrons cannot stay there in material state, because in material form they have expansion.

The second fact is that the remote effect is instantaneous. Instantan means "at the same moment" or "infinitely much faster than the speed of light" or "time does not matter, i.e. does not exist at all".

Operations or processes that require time cannot occur in a non-local space. Nevertheless, something must take place in non-local space, namely the mediation of the interaction between the entangled particles. In the way we define the term process, this mediation is a process that even transmits a kind of information through its mediation work. However, it is a form of information that cannot be read immediately, but must first be decoded.

Let's recap: The non-localspace, as it is shown in connection with quantum entanglement, is a space in which there is neither a distance (spatial extension) nor a time. What exists there is information and processes that do not require time.

In this way the non-localarea differs fundamentally from the local area. Because in the local area all processes need time. Information here always requires an extended carrier, be it material or energetic.

Two spaces with opposite properties "spatial extension - without spatial extension" and "time - without time", cannot be the same physical space. They are two different physical spaces. One is called space-time since Einstein's theory of relativity, but one can also say 4-dim space. The other, namely the non-local space, I have given different names in my books, the two most important of which are "metrics-free vacuum"[21] or "nothing"[22].

Since we want to talk in this writing about a possibly existing "life after life", we can also say "this world" for space-time and "beyond" for the metric-free vacuum.

We should now pause for a moment and become aware of what we have just shown, if not proven:

We have just shown that a physical afterlife exists. So not a spiritual-scientific beyond, but a reality! Because into a spiritual-scientific hereafter at the

---

[21] see e.g. Sedlacek: *Der Widerhall des Urknalls*; Norderstedt (2012); p. 119
[22] see e. g. Wrobel and Sedlacek: *Leben aus Quantenstaub*; Norderstedt (2014); p. 58

end of time something could not possibly enter that was before real physical, but only something that also already before had a pure spiritual-scientific nature. But into the physical beyond things of reality can enter.

Unfortunately, the properties of the physical beyond are the exact opposite of what we know from the space of our imagination, but there is no better way to describe this opposite if we do not want to lose ourselves in the realm of fantasy. However, there is the possibility to imagine the hereafter symbolically (see Fig. 4.4).

There is still much preparatory work ahead of us before we can enter the final phase of our reflections. First of all, it is necessary to clarify the nature of the information that can occur in a metrics-free vacuum. Of these in the next section.

How information containing energy is generated

The question that must be answered at this point of our considerations is: Under what conditions can information exist without a material or energetic carrier?

Because, as we have said above, information in physical reality always needs a carrier. A book without paper, without a screen or without an ebook reader is unthinkable. At best, one can imagine the book memorized in the brain. But the brain would also be a material carrier.

So if information gets along without a material or energetic carrier, then only if it is its own carrier. In this case it must be equivalent to matter or energy. The specific information, which is equivalent to matter or energy, we want to call substance information.

Since Albert Einstein's theory of relativity, we have known that mass and energy are equivalent, that is, they are equal. Equivalence means that one can be transformed into the other.

As experiments in particle physics show, the mass of particles in particle accelerators can be partially converted into energy after collision with other particles. This energy radiates away in the form of photons.

On the other hand, light, i.e. pure energy, has already been converted into matterparticles. Electrons are created from photons.[23]

But are energy and information also equivalent, i.e. the same? If so, then energy could be transformed into substance information and vice versa, from substance information energy resp. material mass equivalent to energy could be produced.

If this is so, there is, from a physical point of view, the possibility of a life after life, otherwise a life after life would at best be limited to the humanistic level of books, media or lectures.

And indeed, energy and substance information are equivalent. To understand why this is so, we must first consider the nature of entropy.

Entropy is the no longer usable part of the energy of a closed system. By dispersion or equalization, entropy increases while the energy remains constant as a whole.

---
[23] cf. Wrobel and Sedlacek: *Leben aus Quantenstaub;* Norderstedt (2014), p. 77 f.

Entropy thus strives for a maximum amount (maximum), which is reached when there is no more usable energy available. In relation to our universe, this leads to heat death, the absolute temperature zero. Nothing stirs then. However, this state only occurs after several hundred billion years.

Entropy was introduced into physics as a fundamental thermodynamic state variable. Of two otherwise identical bodies, the one whose temperature is higher contains more entropy. When two bodies of different temperature are in contact with each other, entropy flows from the warmer to the colder body; thus the temperatures of the two bodies are also equalized. In a closed system where there is no heat or matter exchange with the environment, entropy cannot decrease. However, entropy may occur in the system.

Entropy occurs, for example, when mechanical energy is converted into thermal energy by friction. Since the reversal of this process is not possible, it is also called "energy devaluation".

Fortunately, there is a connection between entropy and information that goes back to the founder of information theory, Claude Shannon. It's valid:

1. Total information of a system = Random information + ordered information.
   If one considers entropy as the amount of random information of a system and ordered information as the information of the usable part of the energy of the system, then the total information of the system is always greater than or equal to entropy.

2. However, the size of the ordered information of a system is basically much smaller than the size of the random information. On closer examination, therefore, the ordered information can be neglected and then it applies:
   Overall information
   = Random information
   = **entropy**
   = no longer usable **energy**.

This last statement means that the total information of a system is equivalent to the part of energy that is no longer usable.

And no matter if the energy is no longer usable in relation to the system it comes from, it is and remains energy and the information about the system is equivalent to it.

By the way, if you would like to have a conversion formula to convert energy into information, you will find it among others in the booklet I published: *"Equivalence of Information and Energy"*.[24]

And another piece of good news that comes as a conclusion from the foregoing:

Every real system generates entropy during its work. Entropy, which is a form of energy, namely that which is no longer usable within the system, thus also generates information equivalent to this energy. The information equivalent to energy we have called substance information above.

Regardless of whether machine or human being, each and everyone generates entropy through work, which as an energy form becomes substance information. Humans create entropy through their lives.

---

[24] Sedlacek, Klaus-Dieter: *Aquivalenz von Information und Energie*; Norderstedt (2009), p. 36 ff.

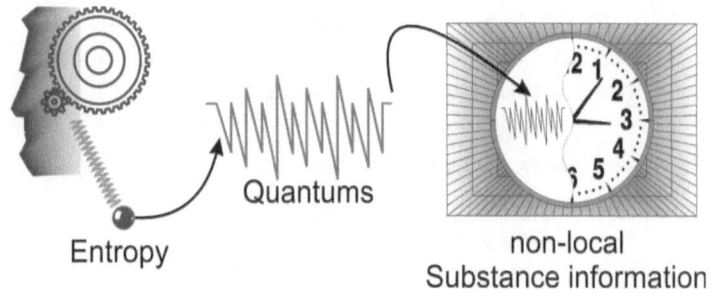

Fig. 4.5: Schematic course of how entropy becomes substance information in the non-local physical beyond. Graphic: Sedlacek

Our consciousness generates entropy through its information-processing processes. And all entropy is a form of energy that is equivalent to information.

Since thinking creates entropy, it is no wonder that our head runs hot during intensive thinking. This way our head also generates a lot of substance information.

Where is the generated information?

A final question that we need to clarify in this chapter is where to find the substance information that generates our life and consciousness.

While work is being done or thought about, heat is generated, namely that heat that is usually no longer usable, which we have called entropy.

Heat is a form of electromagnetic radiation. Light is also known to be a form of electromagnetic radiation. And as we know, light also has particle character, which can be detected in the form of photons. The same applies to heat. As electromagnetic radiation, heat also has particle character.

Elementary particles, which on the one hand show up as small wave-like energy packets, on the other hand as particles, we call quanta.

We already know an important property of quanta from experiments on quantum entanglement. Quantums only assume a real state when they are measured. Before that they have no defined state, or as the physicist says, they are in the state of superposition, i.e. in all possible states at the same time.

Things that have no defined, real state are not things of our space-time. They are not in this world, because things in this world have an extension, cannot have a higher speed than the speed of light and cannot be involved in non-local remote effects. But quanta without a defined real state can have all velocities simultaneously, can be involved in non-local long-range effects and, as long as they are not measured, have no spatial extension. Quantums are obviously not things that belong completely to the world of this world, but at least partially to the non-local beyond.

The only thing that makes quanta, at least to a small extent, things of this world, is the precalculable probability of detecting them by measuring instruments that act like a kind of quantum trap.

But quanta are capricious, because probability does not mean safety. Because there is no certainty of catching quanta where a measuring instrument or other quantum trap is currently located, their behaviour and undefined state cannot be considered to have the same reality character as measurable matter, with mass and energy.

Only when quanta interact with objects on this side of the universe and thus enter a state of mass or measurable energy do they have spatial dimensions and their finite speed, up to the speed of light, can be measured. Then they have only arrived in this world. If they were not in this world before the interaction, then they were in the physical beyond, the non-local physical space without expansion and without time. They were in infinity.

And there in the hereafter, in infinity, they had no mass and no energy, but consisted solely of substance information, the kind of information that is equivalent to energy and therefore also to mass.

Summary of the chapter

If we now look back once more, we have recognized the following in the previous sections: the world is built up and dominated by three fundamentally different, independent manifestations, which nevertheless can enter into close mutual interrelation and connection. In addition, there are other aspects that can give us information about the non-localphysical beyond.

1. Appearance 1: the energetically conditioned material substance, which is dominated by blind regularity, and whose developmental goal is the building of living beings.

2. Appearance 2: The regulatory system that acts primarily on the basis of innate behaviour, which is dominated by unconscious expediency, which organises the substance as a carrier of life, and which is the basis of the living beings that can be carriers of consciousness.

3. Appearance 3: the primarily intuitively recognizing consciousness, whose developmental goal during its connection with body and regulatory system in earth life is a free ethical personality.

4. Further aspect 1: We have found a biophysical theory for extrasensory perception (ASW). Afterwards, the ASW can be explained by the phenomenon of quantum entanglement.

5. Another aspect 2: We have shown that a physical beyond exists with the properties that neither distance nor time exists there. What happens there is information and processes that do not take time.

6. Further aspect 3: Every working process, every process of thinking, every process of consciousness creates entropy in the local this side of our space-time. The entropy then usually changes to the state of substance information, which is equivalent to energy or mass. The substance information remains in the nonlocal physical beyond, if it is not brought into this world by a process of interaction.

With that we have now gained a firm basis from which we can approach the question what the actual aim of these investigations is, namely the question whether there is life after life and what kind of life it will be? She's supposed to keep us busy in the next section.

5. Is there life after the end of time?

Let us once again consider the whole situation. The temporal end is an event with which what we call life seems to have ended. Like an impenetrable darkness it lies over the gate that closes the land of life. Those who pass through them do not return to clear up the darkness, to solve the riddle.

So it is understandable that the doubt arises whether anything can be expected at all after the temporal end, whether life - not only that of the worm, the fly, - but also our life with him is not finally over.

Many believe this and cry out with the poet Heine: A fool waits for an answer! If we nevertheless try to give an answer, it is in the confidence that science has the right to penetrate as deeply as possible into this mystery, and that it will make way for confidence where it has failed so far, or where by its nature it must fail.

The view now that everything is over with the temporal end is the necessary conclusion from the other so-called materialistic view that the living is only a somewhat more complicated manifestation of matter. If this would be the case, then it would, however, finally come to an end with him in the temporal end. Our investigation therefore had to take that materialistic view as its starting point and test its durability.

For this reason we first asked ourselves in the first part about the characteristics of the inanimate and compared life in general. The result was that life necessarily has a special manifestation for itself besides the material-energetic.

Furthermore, in the second part we examined the life of the human being and asked ourselves whether this life in turn was completely absorbed in what we had to call a "regulatory system" according to the previous investigations. The result was that we came across a third special manifestation, consciousness.

These three manifestations are again briefly marked, the following:

1. The substance with its blind-legal energies; its undeniable characteristic is measurability.

2. The system of regulation, its most essential characteristic is the functional and, at the highest level, even rational action based on innate knowledge.

3. The higher consciousness, its characteristic feature in man is creative intuitive cognition and rational and cultural action.

If we want to characterize these three very briefly and strikingly according to their most prominent characteristics, we must say, for example: the "energetic substance", the "innate bio-regulatory system" and the "intuitive higher consciousness". - They are largely independent manifestations, although a step ladder is evident: The substance is lowest, the bio-regulatory system is higher, because it controls the substance; consciousness is highest, because it controls both.

Since the temporal end appears to us as an end, as a cessation, we ask ourselves about the destructibility of these three forms of appearance.

As usual, we are still the clearest on this issue. And especially with regard to the question: "Is it destructible", we are in complete agreement, since it has not only been weighed on the scales, but tested, i.e. for about 240 years.

We have already spoken about this in the first section. The investigation of the substances and the energies has led to a highly important result. There are constant changes in nature, these changes consist of transformations of substances and transformations of energies: But with all these transformations, the inevitable basic law of the entire energetically material nature is revealed: neither mass nor energy is lost, and neither mass nor energy is newly formed. The substance as such is indestructible, what changes in it is only its temporary appearance. And it is the same with the energies.

Now what about the regulatory system and consciousness? Since both are completely different from the fabric, we cannot draw conclusions about them directly from the fabric. Nor can we prove their indestructibility by the same method as this one. This happened with the material by means of the balance, and this is possible because material and energy are measurable. But measurability is precisely a property that the regulatory system and consciousness lack, and it is precisely because of this that both differ fundamentally from the substance, as we have seen.

A different consideration is therefore needed here. If an appearance that builds the world is indestructible, then it seems to be a basic principle of nature. Namely the basic principle that manifestations of nature are only transformed but not lost.

If this basic principle applies generally to manifestations of nature, then it is highly probable that the other two manifestations of the regulatory system and consciousness are also indestructible in principle and are only transformed.

If we compare the laws of nature and information processing, we do indeed find some evidence for this view. Information processing is the area to which consciousness, but also parts of the regulatory system belong. The world is indeed a large, inwardly linked entity. That's their true monism. And this unity is based on the parallelism of their laws in the different fields. From these thoughts one can already suppose that there will be two laws corresponding to the law of conservation of energy or mass: the law of conservation of the processes which control the behaviour of matter, and the law of conservation of information.

But this becomes even more certain when one considers that both manifestations, the regulatory system and consciousness, are the higher ones, dominating that first manifestation. It is reluctant to believe that something destructible should be the dominant principle of something indestructible (matter). If the other two manifestations were destructible, but the mass indestructible, would they be under matter, even though they dominate matter?

The only thing that is usually said against the assumption of the indestructibility of the processes of the regulatory system and consciousness is that when life has escaped from a living being, we no longer perceive anything of it. Yes, in what way should one actually notice anything of him then? That would only be possible by influencing our senses. However, since they themselves are material-energetic, only something material-energetic can have an effect on them. If, however, the processes of the regulatory system as a principle of life are not material-energetic, they cannot influence our senses as such, i.e. they cannot make themselves noticeable to us. This is so self-evident that a word about it can no longer be lost. If you doubt it, you are moving in a fatal circle.

The imperceptibility of the processes of the regulatory system outside the substance is therefore by no means proof of their existence and their indestructibility. On the contrary, it must be said that the possible perceptibility of the regulatory system outside the substance would be a sure proof against its independence. It would then also be - fabric. The appearance of the temporal end itself is therefore by no means proof against the indestructibility of the regulatory system.

But if the regulatory system is indestructible, it is immortal.

"So this is how you think the rose, the snail, the dog is immortal?" they will exclaim in amazement.

According to the preceding argumentation, this question deserves a "yes", of course with the addition: In the sense of indestructibility and immortality discussed so far, i.e. the regulatory system that has been effective in the rose, in the snail, in the dog is not lost, but continues to exist in some way. That this way must again show individual delimitation is not demanded by the previous evidence. On the contrary, by analogy with the indestructibility of matter and energy, we must initially assume nothing more than that the regulatory system returns undiminished in terms of the qualities that make up its essence to the information-processing processes that flood the world.

All assumptions beyond that, e.g. that the individual delimitation belongs to the qualities that constitute the being, that the individual regulation system of the snail trampled on along the way lives on as the regulation system of an individual snail, would require a completely new proof, which in my opinion is impossible.

What happens if we pull out the rose with the root, throw the snail into boiling water, poison the dog? - So we are faced with the mystery of the temporal end? Yeah, what happens?

The necessary living conditions cease to exist, such severe damage occurs that the regulatory system, which has so far bravely resisted the dangers of life, is no longer able to cope with them; the organization of the protoplasm is so violently torn apart that the connection between the regulatory system and the body is severed, the body is no longer usable to serve the regulatory system, and the regulatory system no longer has a material carrier, it can no longer unfold in the body as before: The processes of the regulatory system leave the body of the rose, the snail, the dog.

"Exhaling His Soul" is a thoroughly useful and justified image, not only for the temporal end of man, but also for plants and animals. And these souls were the information-processing processes of the biological regulatory system. - But what leaves animals and plants can be no more than a general principle of life, which we have just called "regulatory system". As such, it is immortal, not as an individual "rose", not as a particular "snail" individual or as the "dog" Hector; for the regulatory system in these beings has not got beyond the processing of stimuli; its special feature was in the body which it has now left and which now sinks back into the dead substances from which it had emerged.

According to this well-founded view, metaphorically speaking, a stream of life, a stream of information-processing processes, is constantly flowing through the world, just as we believe in the generally widespread physical fields, e.g. the electromagnetic fields, which not only flow through the entire space, but also through all materials. And if we relate this finding to what we have experienced in the previous chapter about the entanglement phenomenon, we can assume that the life processes as well as those processes controlling the entanglement phenomenon are to be found in the non-local physical beyond.

Just as the physical fields in thermal, light and electrical phenomena can only appear noticeably where the corresponding conditions are present, so too the information-processing processes of the regulatory system can only express themselves individually where the necessary conditions are present for them, i.e. in the organized substance, in the protoplasm, in the egg cell, which the regulatory system then builds up into an individually formed body.

And just as the physical fields exist outside and without those heat, light and electrical phenomena, so does the information processing principle of the regulatory system as a general, undifferentiated, unconscious manifestation.

Indeed, nothing forces us to go further and assume that rose, snail, dog should continue to live as such after death; but we will have to acknowledge the immortality of that unconscious information processing principle, like it or not.

The temporal end is thus the return of the information-processing life principle to the undifferentiated state from which it had its beginning in the egg cell.

If we want to be consistent and think, we have no reason to make a difference within living beings with regard to the principle of life, i.e. the regulatory system, to assert its indestructibility in humans, for example, but to deny it in animals or plants. Whether then, with regard to its independent existence after the end of time, a certain differentiation still takes place, depending on whether the organization of the body during life was low or high, must be left open, also in this respect we want to acknowledge the possibility and remember that in the field of natural sciences, it is not advisable to use the word "impossible" as long as fundamental laws of nature are not violated. For even the instantaneous long-distance effect of quantum entanglement does not violate the law that there is no speed greater than the speed of light, despite the originally different view of physicists, if local space-time is supplemented by the existence of a non-local space.

The survival of human consciousness

But what about the human being? In view of ourselves, we asked the question whether there is life after life, and we are not thinking of the cessation of that physical life controlled by regulatory processes that we share with plants and animals, but of the spiritual life of our consciousness. In fact, only this one has a true interest in us, the I, which is personally peculiar to us next to our fellow human beings and which is something quite different from the species and genus life in which animals and plants exhaust themselves.

Let us be quite honest: our possible continued existence after the temporal end as a general principle controlling unconscious life processes, which we have been talking about so far, has not the slightest interest for us; because this would basically be the temporal end of what must be of value to us as human beings (not only as living beings) alone, the ego, the personality.

If the mystery of our temporal end was merely dissolved in the separation of an unconscious innate regulation from the body that it had built up, organized and used for a time, then this view can indeed hardly be more satisfying than the return of the body to the blind and dead energies of the substance. The existence after the end of time would then be a dreamlike twilight at best. But such feelings and wishes are of course irrelevant. But as now already our life on earth undoubtedly represents a higher stage, so in that case the end of time would be a sinking down to a lower stage. But this is a very difficult thought to carry out.

For this reason alone we hope that the mystery of our temporal end will not be the same as the mystery of the death of a plant or even of the higher animals.

But now we have already recognized in our second passage that life in man has experienced a very special form, because a special, third manifestation is revealed in man, namely consciousness. From him, what we said above applies.

So we repeat: Consciousness is above the appearance of the substance and the regulatory system because it undoubtedly guides both, it is sufficient, for example, to point to the hypnotic experiment with the stamp paper as a train plaster. But if now the lower, the controlled and guiding principle (the substance) is indestructible, as this is a fixed result of natural science, then it would be an absurdity to believe that the higher and guiding principle, the consciousness, is transient and should cease with the end of time, while the bio-regulation system and the substance continue to exist.

From these thoughts we can conclude the indestructibility, i.e. immortality of consciousness with a certainty that is underpinned by what we know about the nature of consciousness and the non-local physical beyond.

The secret of our temporal end is therefore afterwards in the detachment of consciousness from the body and in the independence of consciousness, which until then was bound by the body and hindered in the full development of its own powers.

As we know, consciousness is an information-processing process to which specific criteria apply so that we can regard it as consciousness. Processes are the connection of control information with a material carrier that executes the process. This means that control information is also information, albeit a type of information that is not itself equivalent to energy, but only in connection with a physical process.[25]

We know from information that it is not bound to a certain carrier. Just like an e-mail, which on the one hand has an electromagnetic carrier, but on the other hand can be transferred to paper, the control information of the consciousness processes and any inform-

---

[25] cf. Sedlacek: *Die letzten Ursachen*; Norderstedt (2015), pp. 183 f., or Sedlacek: *Kleines Wörterbuch der Natur-Philosophie*, Norderstedt (2016), keyword: information.

ation at all that is related to consciousness can of course also be transferred to another carrier.

As carriers of substance information we have already identified the non-local physical beyond. Every entropy that has occurred during life and thus also during conscious thinking and feeling has already been stored in the hereafter in the form of substance information. Even if, due to some interaction, one or the other unit of information is converted back into energy, it would not be lost, because energy is indestructible.

And we have also already talked about the fact that processes can be stored in the non-local afterlife. This is to be regarded as a matter of course, because the natural behaviour of physical objects is not stored anywhere else. How else wouldparticles such as photons know how to behave in Professor Zeilinger's entanglement experiments? Of course there are laws of nature behind the behaviour. But laws of nature are processes[26] that must be stored somewhere. In the photon itself, the quite extensive process information that controls the photon behaviour is certainly not stored. Especially not because there are photons whose energy content is so low that the energy is just enough for the existence of the photon, but for nothing else. So it is probably the case that also consciousness processes are either already stored in the hereafter or are transferred there.

Herewith we have admittedly only generally explained the further existence of consciousness. But that cannot satisfy us yet. We want to lift the veil that lies over the mystery of the temporal end a little further, and above all we want to know what kind of

---

[26] cf. Schlick, Moritz and Sedlacek, K.-D.:*Naturphilosophie: Das Wesen von Naturgesetzen und die Erklärung des Lebens,* Norderstedt (2015), p. 43 ff.

continued existence of consciousness will be, which is now already highly probable, namely whether our consciousness continues to exist as a personality.

When life has disappeared from an organism, then the energies of the substance are left to themselves without any guidance, they no longer show that peculiar purposeful direction as they did during life, but only the blind law that prevails in the whole inanimate world. The process of decomposition leads to chemical equilibrium, in which the substances remain rigid and inert, unless they are reintroduced as nourishment to the operation of life, which, as we see, is characterized by constant disturbance of the chemical balance.

With consciousness, it could be so: If it has detached itself from the body in the transition to the non-localphysical area, it must then continue to show those properties which already distinguished it during its life before the substance. We now want to remember the same as we got to know in the second section. We will find an important argument for the survival of consciousness, but then we will also have an idea of what kind of survival will be.

If we take into account the essential qualities and abilities of consciousness (i.e. receptiveness to suggestion, inductive thinking, deductive thinking, memory, thought transmission, inward looking, moral will, ethical compassion), we can distinguish the following three groups.

1. abilities which belong exclusively to brain activity, such as inductive thinking

2. abilities which belong to the brain and pure mental activity together, namely deductive thinking and memory.

3. abilities which belong exclusively to the higher conscious activity. Intuition, culture and perhaps even extrasensory perception belong to it.

Now what do these abilities of consciousness mean? We will find an answer to this question and a good argument for living on after the end of time, albeit a natural-philosophical one, when we become clear about the meaning of life.

On the one hand there is the opinion of those "practical materialists" who see the meaning of life exclusively in the fact that one enjoys bodily pleasures here on earth for a time. For such a view of the meaning of life there is no mystery of the temporal end. Because for them the temporal end is the conclusion of the earthly joys and the final destruction by bacteria and worms.

On the other hand, we can point out that the goal of development controlled by the principles of cosmological, chemical and biological evolution, which we can recognize, was the formation of living beings as carriers of the brain. And that these living beings evolved into carriers of that wonderful higher consciousness that is inherent in us humans.

The development of mankind, however, stands out from the purely biological objective, which can be described as the best possible reproduction. There are recognizable aims towards a realm of ethical personalities.

If personality is the spiritual self of man with abilities beyond the purely biological, it becomes an ethical personality through moral will as well as through a sense of responsibility towards the human community. The ego of the ethical personality also determines where the processing of stimuli and the instinctive life demands a different direction, contrary to the ethical motive.

Such ethical personalities cannot grow up on the ground of matter alone, but only in the high air of consciousness; but also of only one consciousness as a powerful and independent manifestation, which is superior to matter and the developmentally innate behaviors, but not secondary to them. Here above all the dominating nature of consciousness must be fully shown to advantage. But such an independent manifestation of consciousness is in contrast to those monistic worldviews that deny this independence of consciousness from matter by either letting consciousness merge with matter or letting matter merge with consciousness.

It is precisely from the view of consciousness as an independent manifestation that we now put forward the hypothesis that the purpose and goal of human life is the development of an ethical personality and thus the development of consciousness to full independence and freedom and, with regard to the whole ofhumanity, the development of a kingdom of such personalities.

In the light of this view, the world appears to us as a large, unified, meaningful system with a high development goal, while without it the world would remain random, torn apart and probably senseless.

That goal of human life demands development of consciousness, be it of the individual human being or of all humanity, and for this highly significant purpose humanity is inserted into the whole earth. This development of man is connected with the uplifting of culture and civilization. These in turn are not possible without subjugating nature and exploiting its forces.

But if we ask about the means that were necessary for this, it is clear that it had to be inductive thinking, i.e. slow and laborious investigation and subjugation of the world. An existence of effortless enjoyment would never and never ever have educated mankind to the present powerful culture, but would have caused it to live a limp life.

The ability of inductive thinking and processing of sensual experiences corresponds exactly to the environment in which mankind grew up and in which every single person still grows up today. It would, however, be completely useless for an existence outside this environment.

It is therefore understandable why precisely the brain activity of consciousness, which is closely connected with the sensory organs necessary for the registration of our environment, has an effect in inductive thinking, and it is also natural that this kind of thinking ceases with the end of time, which one is generally mistakenly inclined to regard as the highest. Materialism and all otherphilosophies of this world are right to do so, but they live in the fatal delusion that this way of thinking and recognizing is the only and highest.

Let us note, then, that the subjugation of nature to culture as the free creation of man is a high goal of humanity; but this one goal can only be a means to a higher end. We all feel this very clearly: we speak of "hollow culture" and by this we mean a culture which, although it possesses great external splendour, unfolds in power and pomp, yet lacks inner content and thus true value.

The subjugation of the world alone is not enough; the activity of man's consciousness leading to it is great and beautiful, and has already produced many wonderful fruits; but it is only to form the basis upon which the ethical personality can grow. This requires mastery of the instinctive inborn behavior, restraint of passions, establishment and maintenance of moral balance, acting in an ethical spirit.

Here, too, the intellect, i.e. sensory experiences and their inductive-deductive processing, must play the leading role in healthy development. And in this direction lies the importance of some other abilities of human consciousness.

If a person has a lower power of deduction and a lower degree of memory or recollection in day-consciousness, this is obviously - and experience confirms it - sufficient for the earthly development of consciousness.

Deduction consists of logically consistent thinking and reasoning. It is the processing of the material given by brain activity and induction. Undoubtedly, the combination of inductive and deductive thinking during life on earth is the necessary basis for the spiritual education of man. The inductive experiences of day-consciousness have a suggestive effect on the mind, dominate its thinking and educate it in certain, manifold directions.

The fact that the deductive power of consciousness is basically a much greater one than it appears in life on earth, indeed that it is perhaps almost perfect, results from the fact that, when brain activity is switched off (e.g. in hypnosis), consciousness draws the necessary conclusions with almost inexorable consistency from what may be completely false material presented to it by foreign suggestion.

During daytime consciousness, i.e. full brain activity, this wonderful ability of consciousness does not find its full activity.

It is similar with the memory. In life on earth it is active as perfected "memory", which in its way is already the responsibility of higher mammals with a perfect brain, so it is certainly an activity of the biological regulation system that takes place in the brain. For the earthly development of the human spirit this kind of day memory is perfectly sufficient. We alone have seen from certain phenomena of mental life, which are connected with the shutting down of the brain, that consciousness itself has an almost perfect memory and recall, which cannot be explained by the activity of the nerve cells. And also this perfect memory finds no adequate activity in life on earth.

But as far as now the actually characteristic abilities, which are the sole responsibility of consciousness, are concerned, we must first of all say of intuition, the inner looking, that it is only very rarely effective, actually only under conditions that are abnormal for life on earth. In fact, even intuition would not be very appropriate for the purposes of daily life, because in it, as we saw, for the development of the human race it is necessary to conquer the world, which in reality can only be done by the laborious path of induction.

On the other hand, we see that in certain cases the ability of intuition does extend into life on earth. These are then great scientific discoveries, artistic creation and spiritual experience, and we call the people concerned "geniuses". Very significantly, one has spoken of the "madness of the genius", in which there is indeed a core of truth, if one considers "madness" to be an elimination of the brain. For in the deeds of the genius, parts of the brain's activity are really more or less excluded in favour of intuition, of inner vision.

There is no doubt, however, that such acts of intuitive genius have often enough brought humanity forward in leaps and bounds in its development.

In any case, however, we have in intuition a capacity of human consciousness that is by no means fully active in life on earth, but only exceptionally.

As another ability of consciousness we finally got to know the extrasensory perception. Even though consciousness may use the brain as a receiving and conducting organ, similar to the so-called antennas in radio transmissions, it is certainly not a brain activity requiring sensory experience, as this is more or less switched off.

The same applies to extrasensory perception as to intuition: it only comes into effect exceptionally in earthly life.

To sum up, we must say that human consciousness possesses three abilities above all, namely intuition, perfect memory and possibly extrasensory perception, which play only a very small and temporary role in life on earth, although they are present.

But this is a most striking phenomenon.

It wants to seem almost incomprehensible to us, especially in view of the law of thrift, which otherwise dominates nature in such a striking way and according to which no energy is wasted. So here it would then really be a waste in the spiritual field. What is to be made of it, if such significant abilities remain largely unused and perish with the end of time?

Fig. 5.1: Wreath of bristle-shaped hairs above the ovary, which have no function or purpose during the flowering period Only after the end of the flowering, something develops from it. Photo CC0: Nize

There is a way to solve this problem, and we find this way through biology.

Analogies and laws of nature

Biology teaches us the established fact that, as a rule, there is no organ in a living being that does not serve a specific need and perform a specific function. This goes so far that it is highly probable that we can claim or even assume that an organ whose function and significance we do not yet know has had such a task in the past, or that it will have one later.

We are particularly interested in such cases where organs are present that will only develop fully and function in a future stage of development. Two examples may illustrate this, one botanical and one zoological.

The flowers of the commonly known dandelion have a wreath of bristle-shaped hairs above the ovary, which represent the otherwise quite differently formed calyx. In vain we ask ourselves what this structure means, we discover no function or purpose during the flowering period. But when this is over and the blossom as such has ceased to exist, that is to say, when the temporal end has come about, so to speak -- then that structure suddenly begins to unfold, it grows out of the nascent fruit into that wonderfully delicate parachute that everyone knows, and with which the fruit, when one blows against the spherical fruit stand, imitating the wind, continues to fly gently floating. Now we can see it: those bristles, in their complete unfolding, represent a flying apparatus with which the ripe fruit is able to spread far and wide, as can be seen in every location of the dandelion.

And a zoological example. The female mammals have glands on the abdominal side without secretion and without any detectable function until they have young. But then they suddenly start to secrete milk, with which the young are fed. So here again the complete development and functioning of an organ in a later stage of existence.

We can nevertheless regard it as an established fact that every organ, every ability of a living being has a certain function, and if an organ is still untrained and undeveloped, we will be able to say with certainty that it will in any case have to perform a certain function again in a later stage of existence of the being concerned.

What applies in the field of biological regulation, as just shown, can apply in an analogous way to consciousness. The parallelism of the two areas, which has already been emphasised, should be pointed out here once again. If we find abilities of consciousness that cannot be fully or almost not at all activated during life on earth, this can be seen as a strong indication of future use.

It is highly probable that we can now formulate the hypothesis that is the culmination of our entire investigation: An ability of consciousness that is not usually activated in life must be given the opportunity for full and meaningful activity in the future.

But since life on earth does not offer the possibility at all to exercise certain abilities of consciousness, it follows that life on earth will be followed by another form of existence of consciousness without the body that hinders it, without the brain that is designed to develop through sensory experience.

From significant thought processes we have arrived at a well-founded hypothesis for a life after the end of time, i.e. for the imperishability of our consciousness. Certainly, the hypothesis is based on analogies, and it cannot be denied that analogies can never have the probative force of physical experiments, but in the field under investigation, can one expect anything more than analogous proof?

Yeah, a little more than that. And this more I will present here. There are physical laws that support our hypothesis.

The question of what can exist in the physical beyond has already been answered above. We have come to the conclusion that in the hereafter there are information units and processes. The information reaches the hereafter already during the existence in this world, simply because all processes in this world create entropy and this entropy finally lands in the hereafter as substance information via the path of quantum properties. Since our consciousness also generates entropy when working in this world, the information belonging to consciousness is ultimately stored in the hereafter.

But the answer is not enough for us. Because if the information from our consciousness is stored for years in the hereafter, it could be that the stored information does not form a coherent unit, but is scattered somewhere in an incoherent manner. But a scattered state without connection of the information units would be equivalent to the dissolution of consciousness after the end of time.

I don't think we'd really like that. We rather expect that the entirety of the information of our consciousness forms a coherent unit in the hereafter as well. Only this totality would correspond to our personality, our identity.

In order to answer this question, we first have to understand how and in what way information is con-

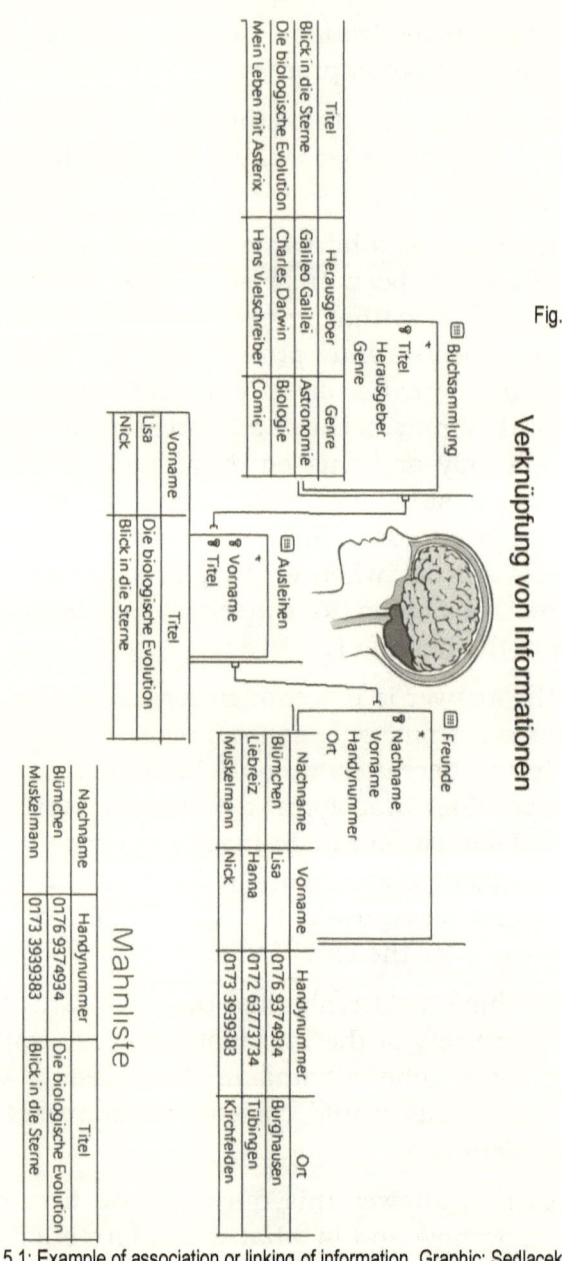

5.1: Example of association or linking of information. Graphic: Sedlacek

nected in the first place and can thus form a uniform whole.

To answer this question I have designed a graphic with a simple example (Fig. 5.1). The example is about information about friends, a book collection and borrowed books.

The friends are stored in the head with last name, first name, place and even mobile phone number, because our example person has a good memory for numbers. The connection results from the identification as "friends" and so it is also stored in the brain.

When someone hears the keyword "friends", his brain starts working, and he can immediately list a few friends even by last name.

In the same way, the data of the book collection are linked in the brain. By specifying the keyword "book collection", a respondent can immediately list a few books from his collection.

If now friend Lisa comes and borrows the book "The biological evolution", then a new association (= link) is stored in the brain. It is the association "Book Lending - Lisa - The Biological Evolution". Interestingly, the book lending now indirectly links the friends with the book collection. Because if you search for the book "The Biological Evolution" a few weeks later, you will come from your book collection stored in the brain to the information who has borrowed something, although this knowledge is stored in a completely different place in the brain.

The brain brings the borrower "Lisa" into consciousness, because "The biological evolution" is linked to Lisa.

Now another thought process begins. Lisa is to be warned by telephone to return the book. Although no mobile phone numbers are assigned to the borrowings, the brain finds the information it is looking for via the link "borrowing - Lisa - friends - Lisa - mobile number". Lisa can now be called and reminded of the return.

The individual pieces of information can be scattered all over the brain. Thinking processes find what they are looking for simply by following the stored links. In this way, the mobile phone number is associated with the borrower of the book.

This system only works if, in addition to the stored links, there are processes that retrieve the associated information and, if this information cannot be retrieved easily, the links can be traced and the searched for can be found in other storage locations.

But because a link is also information, the link information is stored permanently in the hereafter just like other information, as already discussed. Therefore it is indifferent if the information is apparently scattered in the hereafter. Through the links everything forms a unit or totality. And this totality of information makes up the personality or identity of a person.

There is only one point left to say from a physical point of view. The information is dead if there are no processes that use the information. Only processes can make associations or find what is sought in the jumble of linked information.

But what processes might these be in the hereafter that bring the information to life and turn dead information into a living consciousness or personality?

Biological processes in this world consist on the one hand of information that controls the processes, but on the other hand of a carrier that is a small material biological machine. Without matter no processes take place in this world. In addition, it needs energy to drive the biological machine.

But what about the hereafter? There can be neither matter nor energy. Is the beyond a gigantic information grave into which everything that exists falls until the heat death of the universe and then everything is over?

No, it's not like that. There is a process that underlies all other processes in the universe and that always works, even when the heat death of the universe has occurred. This process is called fluctuation.

Fluctuation refers to a random quantumphysical process that causes a spontaneous change of circumstances and states.

When a physicist uses the word quantum leap in his mouth, the phenomenon is due to fluctuation. When a physicist talks about the tunnel effect, in which a quantum overcomes a barrier that it actually cannot overcome due to a lack of energy, then the fluctuation is behind it. When a quantum decides on a real state, for example in an experiment on quantum entanglement, then the fluctuation is behind it. And when particle physicists create particles or even electrons out of nothing with the help of energy input, it was not Harry Potter who was at work, it was fluctuation that, within the framework of quantumphysical laws, created matter.

And fluctuation is responsible for the fact that even at absolute zero temperature, when nothing should be moving, there is still bubbling in the vacuum and particles are created or die away.

Fluctuation is pure coincidence, not calculable in any way, it is the movement and the motor for everything that happens in our universe. Only on the macro level, when a huge number of smallest coincidences come together, then the laws of classical physics emerge as statistical mean values of coincidences. - By the way, there is a small writing by Moritz Cantor, in which he describes how the law arises by chance.[27]

Fluctuation is the basic process that also occurs in the non-local physical beyond and serves as a drive for control information of processes stored there. From the hereafter, the fluctuation drives natural processes according to which the smallest particles, the quanta, act. As already mentioned above, it has not yet been possible to discover any storage space in quanta for the control information that controls the course of natural processes.

And so it is also the fluctuation that can breathe life into consciousness processes in the hereafter, and which, if there is actually a life after life, provides for the liveliness of life in the hereafter.

There is, however, one point to bear in mind with this scientific theory of the hereafter: theories are never the absolute truth, but are always put to the test. But as long as no experimental facts speak against it, we can confidently rely on its accuracy.

This was the presentation from the point of view of the physical side. I think those who ask for more ask for the impossible, even nonsense.

And so, with our logical and scientific thinking we have penetrated so far into the mystery of the temporal end that we may say An independent continued existence of consciousness after the temporal end outside the body becomes highly probable due to bio-

---

[27] Cantor, Moritz and Sedlacek, K.-D.: *Das Gesetz im Zufall. Wie sich verborgene Gesetzlichkeit manifestiert*, Norderstedt (2016)

logical analogies, physical laws and the existing facts of nature and consciousness life.

The secret of the temporal end, therefore, is that it means the return of the body to the inanimate energies of the matter, but also the release of the shackles of consciousness from the limitations of physicality. It is the liberation of consciousness from the bonds of its hitherto necessary evolutionary development and it is the birth of its actual independent existence.

If in the earthly birth the body is born to independence, then at the temporal end it is the consciousness that is born to independence.

Hours of birth are moments of transition. They are connected with the detachment from a previously close community and usually also with struggle and inconvenience. This is how it is when a life that has previously grown together with the mother's organism is torn away from it. Can it be different when the consciousness frees itself from the body in which it has grown to perfection so far?

So the transition into the non-local realm, i.e. into the physical beyond, where the physical bonds are loosened and finally completely released, is not always, but mostly connected with a fight.

What we fear is the transitional struggle of the body; but we should welcome the birth of consciousness as the achievement of a necessary stage of our existence, as the beginning of a higher life. Inconvenience may well come with the dissolution of the body; but not with the liberation of consciousness. On the contrary, this will rather be a moment of bliss. And indeed, all the signs speak for that.

Except for cases where people have been surprised by a sudden, violent end, it is observed time and again that after the body has fought, there is peace. Improvement of the suffering some time before the end of time is a very common phenomenon, which often enough feigns the beginning of recovery for the relatives. And almost proverbial is the peaceful smile on the face of those who have fallen asleep after suffering. After rescue, drowning and falling persons tell that their consciousness was filled with cheerful images of their past life.

These are all indications that when the resistance of the body, of matter, has ended, the consciousness now begins to move freely, to shake off its wings, to cast off its shackles and to rise into a more blissful environment, into a higher being.

The organization of life after life

And now the question that will be close to everyone's heart: How can life be shaped after life? What form of existence will the disembodied consciousness have?

From the outset it is self-evident that we cannot even begin to form an accurate idea of this non-bodily existence, because as long as we live in the body, we necessarily lack all the prerequisites for it.

The story of those two monks will always keep its deep truth: They had talked a lot about the hereafter and imagined it in their own way. They promised each other on the most solemn of occasions that the first to die would appear to the survivor in his sleep to tell him about the hereafter, in a single Latin word.

-- If the deceased said "taliter" (i.e. the same), the survivor was to know that in the hereafter it would be as both had imagined it; if the survivor said "aliter" (i.e. different), the survivor knew that in the hereafter it would be different than they had imagined. One of the two monks died and appeared to the other one in his sleep: but he said: "totaliter aliter" i.e. completely different.

So it will also be for us, when we awaken from the body to the liberated life of consciousness.

After all, the investigations of consciousness have revealed so much and so significant about its abilities that we must at least be able to find the foundations of a disembodied existence.

The first question will be whether consciousness will be in complete inactivity after the temporal end, as some people might imagine. Further reflection will reveal the intolerability of this idea. In fact, this would basically be tantamount to destruction.

If we speak of a "life after life", we must also be serious about the term "life". "Life" is work and creation and activity.

For us, the question arises whether and how consciousness can be active in an existence that has to do without a body like the one we have now. For the answer to this question, what we have recognized as abilities of consciousness must be decisive.

We have seen that consciousness possesses abilities that it can never fully develop and use in life on earth. We drew the conclusion that there must be a time for the consciousness to develop and use these abilities once again, i.e. the conclusion of a life after life.

Now we add that this life after life will consist in a free development, effectiveness and activity of those very abilities of consciousness.

We have also seen that the nature of consciousness is expressed in those three basic functions of thinking, wanting and feeling. This will also be the case in the state of the hereafter; but then not depending on the sense organs and the brain.

With regard to thinking, this means that the way of recognizing and experiencing will then no longer be the laboriouslyinductive one; because the basis for this will be missing: the ideas gained through the senses. The basis of cognition will now be rather the inner looking, the extrasensory perception and the intuition, as well as the effortless viewing and reception of the worldly information and the worldly being.

But it will not be a merging into an all-spirit as in Buddhism. The other abilities of the consciousness certainly speak against this. One of them is above all an almost perfect memory. This will ensure that consciousness will gain knowledge from life on earth. Intuition and memory will provide further material, which will now be processed by consciousness according to its ability to deduce.

But to the acquisition from life on earth also belongs what we have worked out in our will and feelings, and also this will be preserved for us by means of the ability to remember. And when we consider how many unfinished things in this direction often enough seem to perish with a human life here, how many unbalanced opposites, how many broken and unresolved existences and how many torn off possibilities of development the earthly existence has, --

who would then deny the possibility that we can still expect a balancing of opposites, a clarification of the unresolved, in short, a further development in the hereafter! But it is clear that this further development must be influenced by the result of life on earth, by what we have worked out in this development school.

One question that may be on the tip of everyone's tongue is, will we find those who were close to us in life again? I don't see why not. For also their friendship and love belongs to the acquisition of this world, which must follow us into the hereafter according to our memory. But if we ask how then an intercourse of beings must still be possible there, then it will certainly no longer take place physically, i.e. through eye, ear, etc.; but let us remember that consciousness already here possesses a capacity that permits intercourse with other beings beyond the senses: The extrasensory perception. It will be the means of communication of the disembodied entities, in that the thoughts will pass between them as free forces.

In this way, thinking, wanting and feeling will accompany us into the future existence, even if in a different form than before, so that what we must already now consider our most valuable possession, our personality, will also be preserved. Accordingly, our survival will not be an impersonal merging with an All-Spirit, but a personal survival.

We have spoken of the "this world" and the "hereafter", there is still a fundamental error to be faced, which has woven itself around the mystery of the temporal end.

Space and time are closely connected with sensory experiences and therefore with brain consciousness as "forms" of its viewing and thinking. They belong to the inductive conception and comprehension of the material world. The "hereafter" must by no means be understood as a space for visualization, as we experience it in this world. For the world of consciousness, sensual experiences are non-representational. Thus the problem of temporal eternity and spatial infinity collapses. Eternity and infinity do indeed have a meaning, but not as "eternal time" and "infinite space", but as timelessness and spacelessness.

But if there is no space and time for pure consciousness, that is to say after the end of time, then we must also not conceive of the "hereafter" as already mentioned several times. In terms of consciousness, it is not a place of contemplation, but a form of existence. It should better be "on this side of the body" and "beyond the body".

Summary of the chapter

We are at the end of our investigation and now we will briefly summarize once again: We proceeded from an investigation of the inanimate and the animate. The result was that we found a fundamental difference between the two: besides the appearance of the substance, there is a second appearance as a principle of life, the "soul", which we identified as the bioregulatory system of the body.

The study of the nature of what we call the "consciousness" of the human being has led us to discover a further manifestation with special abilities that differ from those of the substance and the regulatory system.

But the real question that concerns us is whether there is a life after life? And to answer them, we examined the indestructibility of these three manifestations. The same has been proven for the substance, so it seems very likely for soul and consciousness. But security offered us another consideration in terms of consciousness.

The investigation of consciousness revealed to us some of its abilities (intuition, perfect memory and extrasensory perception), which play only a very small or no role at all in life on earth.

In addition, we discovered relationships to phenomena of quantum physics. These provided us with the physical confirmation of our considerations.

Biology alone shows us numerous examples that organs do not yet play a role at a certain stage of life, but then become functional at a later stage of life. We may assume that this will apply to a higher degree of the now still latent abilities of consciousness, i.e. another stage of being will follow earth life, on which those abilities will come to full development and effectiveness.

And from these abilities we will then be able to get a rough idea of what life will be like after life.

We readily admit that with the last touched we enter an area where further conclusions are vulnerable; for as long as we stand in earthly life we will never be able to form a real picture of a disembodied existence of consciousness. As that monk said, "Totaliter aliter (totally different)", we always want to remember that.

It also touches on the area in which scientific thinking by its very nature must have its limits.

My task should be to lead the reader on the scientific path as far as possible, and to examine whether we can gain a conviction from him or her for life after life.

I hope that this path, which may seem sober, has already brought many a thing to some people: especially confidence based on facts.

And the second is: a drive to shape our existence in such a way that life as a development to an ethical personality demands it, to overcome the matter within us and to live ourselves completely into a consciousness existence that already begins here on earth and projects into the light of timelessness.

How are we to face the end of time, - it comes closer to us with every swing of the pendulum. Let us surrender ourselves to the equanimity of the Stoic, who says: "I cannot know anything about what is coming now, after all, and penetrates from the depths of his inner being: Suppose he does?

Or do we want to approach the great hour with confidence, knowing that behind the gate of the temporal end the light of a brighter life is flooding into the liberated consciousness?

Is the temporal end to defeat us or do we want to defeat it? - But who is it that beats it? For example, the one who expects his prank indifferently, or the one who faces the transition into the timeless realm with inner confidence?

We want to grow out of the darkness of the earth and matter and up to the light of knowledge, which reveals itself to those people who gain something important for themselves from these three stages of joint exploration of life after life:

> Joy about the liberation of consciousness from the shackles of time.

6. Literature

Auerbach, Felix: *Raum und Zeit, Materie und Energie;* Dürr'sche Buchhandlung, Leipzig (1921)

Bekenstein, J.D.: *Phys. Rev. D7* (1973) 2333 und *Phys. Rev. D23* (1981) 278

Bennet, Charles H.: *Maxwells Dämon;* in: Spektrum der Wissenschaft, Heft 1/1988, S. 48-55

Blome, Hans-Joachim u. Zaun, Harald: *Der Urknall – Anfang und Zukunft des Universums;* München (2. aktualisierte Auflage 2007)

Born, Max: *Die Relativitätstheorie Einsteins;* Springer, Berlin-Heidelberg-New York (1969)

Bouwmeester D, Pan JW, Mattle K, Eibl M, Weinfurter H, Zeilinger A (1997) Experimental Quantum Teleportation, Nature 390: 575-579

Churchland, Paul M.: *Die Seelenmaschine. Eine philosophische Reise ins Gehirn;* Spektrum, Heidelberg (2001)

Davis, Paul: *Der Plan Gottes. Die Rätsel unserer Existenz und die Wissenschaft;* Insel Verlag (1996)

Dawkins R (2003): A Devil's Chaplain: Reflections on Hope, Lies, Science, and Love. Houghton Mifflin 2003, ISBN 0-618-33540-4

Dawkins, Marian Stamp: *Die Entdeckung des tierischen Bewusstseins;* Spektrum, Heidelberg (1994)

Dennet, Daniel C.: *Spielarten des Geistes;* Bertelsmann, München (1999)

DIN 19226 Teil 1, Deutsche Elektrotechnische Kommission im DIN und VDE (DKE) Februar 1994

Driesch, Hans: *Metaphysik;* Hirt, Breslau (1924)

Dubislav, Walter: *Naturphilosophie;* Junker und Dünnhaupt, Berlin (1933)

Dürr, Hans-Peter, Hrsg.: *Physik und Transzendenz,*; Scherz (1989)

Ebeling W, Feistel R (1982) Physik der Selbstorganisation und Evolution. Akademie Verlag Berlin, S. 83 ff

Einstein, Albert: *Über die spezielle und die allgemeine Relativitätstheorie;* Vieweg+Sohn, Braunschweig (1973)

Einstein, Albert: *Zur Elektrodynamik bewegter Körper;* In: Annalen der Physik. 322, Nr. 10, 1905, S. 891-921

Feynman, Richard: *Vorlesungen über Physik;* Band II, Oldenburg (2007), Kap. 15-4.

Flammarion: *Rätsel des Seelenlebens;* Stuttgart, J. Hoffmann (1909)

Froböse, Rolf: *Die geheime Physik des Zufalls;* Norderstedt (2008)

Görnitz, B & Th.: *Der kreative Kosmos – Geist und Materie aus Quanteninformation;* Spektrum, Heidelberg (2007)

Görnitz, Th. Graudenz, D., Weizsäcker, C.F.v.: *Quantum Field Theory of Binary Alternatives;* Intern. J. Theoret. Phys. 31 (1992) 1929-1959

Goswami, Amit: *Die schöpferische Evolution. Zwischen Gottesglaube und Darwinismus;* Lüchow, Stuttgart (2009), S. 31 f.

Gould, James L. & Gould, Carol Grant: Bewusstsein *bei Tieren;* Spektrum, Heidelberg (1997)

Griffin, D. R.: *Wie Tiere denken;* dtv, München (1990)

Haeckel, Ernst u. Sedlacek, Klaus-Dieter (Hrsg.): *Die Welträtsel – Gemeinverständliche Studien über monistische Philosophie;* Norderstedt (2009)

Hawking, S. W.: *Particle creation by black holes;* Comm. Math. Phys. 43 (1975) 199-220

Heisenberg, Werner: *Quantentheorie und Philosophie;* Reclam, Stuttgart (2008), S. 43

Herbert, Nick: *Quantenrealität. Jenseits der neuen Physik;* Birkhäuser, Basel (1987)

Hey, Tony u. Walters, Patrick: *Das Quantenuniversum;* Spektrum (1998)

Hofstadter, Douglas R. & Dennet, Daniel C.: *Einsicht ins Ich. Fantasien und Reflexionen über Selbst und Seele;* Klett-Cotta, Stuttgart (1986)

Kanitscheider, Bernulf: *Kosmologie;* Reclam (1991)

Küng, Hans: *Der Anfang aller Dinge: Naturwissenschaft und Religion;* Piper (2005)

Law, Stephen: *Philosophie;* Dorling Kindersley, München (2008)

Lazlo, Ervin: *Holos die Welt der neuen Wissenschaften;* Via Nova (2002)

Monod J (1970): *Zufall und Notwendigkeit.* R. Piper Verlag München 1971, ISBN 3-492 01913-7

Neumann von J (1991): *Die Rechenmaschine und das Gehirn.* R. Oldenbourg Verlag München, ISBN 3-486-45226-6

Penrose R (1995): *Schatten des Geistes. Wege zu einer neuen Physik des Bewusstseins.* Spektrum, Heidelberg/Berlin/Oxford ISBN 3-86025-260-7

Penrose, R.: *The Emperor's New Mind.* Oxford University Press, Oxford (1989; Deutsch: *Computerdenken;* Spektrum, Heidelberg (1991)

Perty, M.: *„Ohne die mystischen Tatsachen keine erschöpfende Psychologie",* Leipzig, E. F. Winter, (1883)

Prigogine I (1980): *Dialog mit der Natur.* R Piper Verlag München 1990, ISBN 3-492-11181-5

Rae, Alastair I.M.: *Quantenphysik: Illusion oder Realität; Reclam,* Stuttgart (1996)

Schlichting HJ (2000) Von der Energieentwertung zur Entropie. Praxis der Naturwissenschaften/ Physik 49(2): 7-11

Schlick, Moritz u. Sedlacek, Klaus-Dieter: *Naturphilosophie. Das Wesen von Naturgesetzen und die Erklärung des Lebens*; Norderstedt (2015)

Schrenck-Notzing, Dr. A. Freiherrn von u. Sedlacek, Klaus-Dieter: *Die Natur Psycho-Physikalischer Phänomene. Erforschung telekinetischer Vorgänge;* Norderstedt (2009)

Schrödinger (1989): *Was ist Leben?* R. Piper GmbH & Co. KG München 1987, ISBN 3-492-11134-3

Sedlacek, Klaus-Dieter: *Äquivalenz von Information und Energie. Auf der Suche nach den Grundbausteinen der Welt;* Norderstedt (2009)

Sedlacek: *Kleines Wörterbuch der Naturphilosophie: 1200 Begriffe, die man kennen sollte, kurz und prägnant;* Norderstedt (2016)

Sedlacek, Klaus-Dieter: *Supervereinigung. Wie aus nichts alles entsteht. Ansatz einer großen einheitlichen Feldtheorie;* Norderstedt (2010)

Sedlacek, Klaus-Dieter: *Synthetisches Bewusstsein. Wie Bewusstsein funktioniert und Roboter damit ausgestattet werden können;* Norderstedt (2011)

Sedlacek, Klaus-Dieter: *Unsterbliches Bewusstsein. Raumzeit-Phänomene, Beweise und Visionen;* Norderstedt (2008)

Sedlacek, Klaus-Dieter: *Der Widerhall des Urknalls;* Norderstedt (2012)

Shannon CE, A Mathematical Theory of Information. In: Bell System Technical Journal. Short Hills N.J. 27.1948, (Juli, Oktober): S. 379–423, 623–656 ISSN 0005-8580

Sharov, Alexander S. u. Novikov, Igor D.: *Edwin Hubble. Der Mann, der den Urknall entdeckte;* Birkhäuser, Basel (1994)

Sperling, Jan: *Untersuchung von H/D-Isotopeneffekten bei der elektrolytischen Wasserspaltung im Hinblick auf eine mögliche Quantenkorrelation;* Dissertation, FU Berlin (1999)

Szilard, Leo: *Über die Entropieverminderung in einem thermodynamischen System bei Eingriffen intelligenter Wesen;* In: Zeitschrift für Physik 1929; 53: 840-856

Tipler, Paul A. Und Mosca, Gene: *Physik für Wissenschaftler und Ingenieure;* 6. Auflage, Spektrum (2009)

Verweyen, J.M.: *Naturphilosophie;* Teubner, Leipzig (1915)

Volkmann, Paul: *Erkenntnistheoretische Grundzüge der Naturwissenschaften;* Teubner, Leipzig (1910)

von Weizsäcker, Carl Friedrich: *Aufbau der Physik;* Hanser, München (1985)

von Weizsäcker, Carl Friedrich: *Die Einheit der Natur;* Hanser, München (1971), S. 269

Wilber, Ken: *Naturwissenschaft und Religion. Die Versöhnung von Wissen und Weisheit;* Fischer, Frankfurt (2010)

Wrobel, Norbert u. Sedlacek, Klaus-Dieter: *Leben aus Quantenstaub*; Norderstedt (2014)

Zeilinger, Anton: *Einsteins Spuk: Teleportation und weitere Mysterien der Quantenphysik*; Goldmann, München (2007)

7. Keyword index

Ability21f., 38, 40, 42, 70f., 75, 77f., 107f., 110f., 116, 125f., 128f.
Alcohol20
Allgeist126f.
Alternative133
Amphibians23
Analogy100, 114, 116f., 122
Anesthesia67
Anesthesia67ff.
Antenna113
Appearance27, 32ff., 37, 47, 53f., 60, 78f., 94ff., 102, 104f., 109, 128f.
appropriate17, 19ff., 23, 25ff., 30, 36ff., 80, 97, 107
Aristotle31
Association42f., 48, 119f.
ASW55, 59, 80, 95
Atom24, 81
Auerbach, Felix: Space and time, matter and energy; Dürr'sche Buchhandlung, Leipzig (1921)132
Awareness2f., 7, 18, 38f., 42, 45ff., 51ff., 60ff., 64ff., 73, 75ff., 83, 91f., 94f., 97ff., 103ff., 116f., 119f., 122ff., 145
Bacterium16
Balance system29
Balance18f., 28f., 107, 111
Basic function126
Basic Law13, 23, 98
Basic material15
Basic module135
Basic principle98
Behaviour7f., 12, 25, 38ff., 45ff., 50, 58ff., 85, 93f., 99, 106, 109, 111
Being127
Being145
Bekenstein132
Bell135
Bennet132
Beyond36, 56, 83, 87f., 93ff., 102, 105f., 117, 120ff., 134
Big Bang132, 136
Bio-Regulation System28, 30ff., 40, 52, 78f., 97, 105, 128
biochemical29f., 65, 69, 82
Biochemists16
biological system29
biological28f., 35, 37, 43, 48, 52, 101, 108, 112, 116, 119ff.
Biology27, 114, 129
biophysics82
Blood19, 82
Body-mind problem34
Body11f., 18f., 21, 29, 32, 34f., 40, 47, 52, 61, 77ff., 82, 90, 95, 101ff., 107, 116, 122ff., 128, 133
Born132
Brain41f., 48f., 53f., 60ff., 65ff., 73, 77ff., 82, 88, 108, 110ff., 116, 119f., 126, 128, 132, 134
Breeding area43
Cancer23
Cantor122
Carrier35ff., 78, 87ff., 94, 101, 105f., 108, 120
Caterpillar39f.
Causality68f., 79
Cause14, 23f., 27, 68, 145
Cell bond29
Cell membrane29
Cell nucleus15
Cell15f., 18, 20, 24, 26f., 29, 33, 35, 38, 41
Central organ41
Chamet63
Change in activity29
Character7, 30, 50ff., 59, 62, 75
Characteristics24, 30, 32, 43ff., 52, 72, 80, 97
chemical12f., 15, 18ff., 24ff., 29f., 32, 36f., 41, 65, 69, 82
Chemicals13, 23
classic13, 55, 83, 85, 122
Clear Trauma 62f.
Cocoon39f.
Colburn74
Collared Flycatcher43ff.
Community43, 109, 123
Compassion107
complementary57f.
Computer134
Condillac63
Condition (interlocking)56
Conflict of objectives18
Control information29, 79, 105f., 122
Control loop28f.
Controller29
Conversion13, 19f., 98
correlates56f., 59
cosmological108
Cosmology134
cosmos133
Culture50, 80, 108, 110
Daily consciousness54, 61f., 65, 70, 73, 76ff., 82f., 111f.
Dandelion115
Darwin133
Dawkins132
Day memory112
Decision18, 43ff., 50, 58
Deduction77, 111, 126
deductive54, 76f., 107f., 111
Demon 132
Dennet132, 134
Descartes63
determined45f., 50
Development2, 31ff., 39, 78f., 94f., 108f., 111ff., 115f., 123,

137

127, 130
Dial43ff., 50f.
Diastase20
Discovery13, 112, 132
disembodied125, 129
Displeasure51f.
Distance81, 85, 87, 95
Dog42f., 50f., 90, 100ff.
Done85
Dreams61ff., 66
Drive50, 52, 109
Ear68, 127
Earth life95, 104, 111ff., 116, 125ff., 129
Ebeling133
Effect14, 20, 23f., 48, 68, 85
Effector29
Einstein8, 55, 59, 68, 81, 87, 89, 132f., 136
electric14f., 29, 84, 102
Electrodynamics133
electromagnetic35, 92, 102, 106
Elektron24, 84, 86, 89, 121
element84
Elementary particles24, 86, 106
energetic26f., 34, 78, 87f., 94, 97ff.
Energy12ff., 17, 23, 25ff, 30, 32f, 36ff, 50, 89ff, 97ff, 104ff, 113, 121, 123, 132, 135, 145
Entropy89ff., 95, 106, 117, 135f.
Environment44f., 110
Environmental condition44f.
Environmental incentive44
Enzyme20, 26, 35
epistemological84
Equivalence of information and energy135
Equivalence89, 91, 135, 145
ethical52, 80, 95, 107ff., 111, 130
Evening peacock butterfly39
Evolution108, 119, 133

evolutionary108, 123
Exchange of information29
Existence132
Expansion84
Experiment8, 16, 45, 47, 50, 56ff., 66, 81f., 85, 89, 93, 105, 117, 121, 132
extrasensory perception55, 59, 64, 70, 77, 80f., 83, 95, 108, 113, 126f., 129
Eye19, 23, 53, 58, 63, 68, 83, 85ff., 123f., 127
Fabric12ff., 17ff., 23ff., 29f., 33ff., 49f., 53, 68, 76, 78ff., 94, 97ff., 102, 104f., 107, 123, 128f.
Feed substance29
Feel48, 51f., 54, 80, 106, 126f.
Fermentation20
Feynman133
Field strength84
Field Theory135
Field84, 102, 135
Final control element29
Flammarion60, 133
Flower115
Fluctuation121f.
Food19, 40, 43f., 46, 107
Form85
Freedom23, 25, 51, 109
Friction90
frogboy133
gene136
Genie74, 110, 112f.
Ghost 136
glycogen19, 26
God132f.
Goethe73
gold136
Görnitz133
Ground13f., 23, 26, 89, 93ff., 98f.
Haeckel133
Handel62
Hawking133

Heat death90, 121
Heat90, 92, 102, 121
Heine96
Heisenberg134
Hofmeister20
Hormone30
Hubble136
Huber70
Human7, 9, 11f., 22, 28, 31, 34, 38f., 41, 43, 47ff., 61f., 66, 68, 74ff., 79, 81f., 84, 91, 97, 101, 103f., 108ff., 124, 126, 128, 131
Hunger50
Hydrogen14f.
Hypnosis66ff., 111
Hypothesis44, 109, 116f.
I7ff., 47f., 52, 60, 64f., 68f., 103f., 108f., 117, 122, 127, 130, 134
Immortality100, 102, 105, 145
inanimate13ff., 17, 19f., 23ff., 36ff., 107, 123
Incineration14
Independence61, 76, 80
Indestructibility98ff., 103, 105, 129
indestructible13, 15, 98ff., 105f.
Induction54, 76f., 79, 111f.
inductive54, 61f., 76, 107f., 110f., 126, 128
Infinity94, 128
information processing system30, 32, 34
information processing30, 32, 34, 37, 42, 46, 48f., 78, 80, 82, 91, 100ff., 105
Information processing34, 44, 82, 99
Information2, 17, 27, 29f., 34ff., 42ff., 49, 71f., 77ff., 86ff., 94f., 99, 105f., 117, 119ff., 126, 135, 145
Inheritance76
innate behaviour38ff., 45

instantaneous 57, 59, 86, 103
Instinct 38, 41
Institution 18f., 21, 23, 28, 33, 35, 38, 40, 48, 54, 79, 101, 103, 107, 114ff., 123, 129
Interaction 12, 27, 38, 79, 81f., 85f., 94f., 106
Interference 28f.
interlaced 56ff., 81f., 85f.
Interlocking 56, 81f., 102, 106
Intuition 62, 73, 75, 77, 80, 108, 112f., 126, 129
Invertase 20
isotope 136
Knowledge 132, 134, 136
Knowledge 83
Kotik 71
Küng 134
law of nature 145
Layer of consciousness 67
Legality 103, 117, 123
Legality 14, 24, 94, 107
Lehmann 72
libidinal 111
Life force 24, 31f.
Life Process 12, 17f., 23f., 40, 102, 104
Life 1ff., 9ff., 15ff., 21, 23ff., 30ff., 36ff., 43ff., 49ff., 57, 78, 80f., 87, 89, 91ff., 94ff., 99ff., 106ff., 112, 116, 120, 122ff., 135f., 145
Light 7, 14, 29f., 42, 55ff., 59, 81, 86, 89, 92ff., 102f., 109, 130f.
Line 26, 113
Liver cell 19f., 26
Living beings 15ff., 25ff., 31ff., 36ff., 40, 71, 77, 94, 99, 103f., 108, 114, 116
Lizard 21
local 8, 34, 36, 55f., 58, 67, 80f., 83, 85ff., 93ff., 102f., 105ff., 122f.
Locomotion 29
Lombroso 60

Lust 51f.
Machine 35, 91, 120f.
Maltase 20
Mammal 112, 115
material 12f., 15, 32, 35, 37, 78, 80, 86ff., 94, 97, 101, 105, 120, 128
materialism 110
Materialist 108
mathematical 84
Mathematics 2, 39
Matter 12ff., 17, 30, 49, 52, 80, 89f., 93, 96, 99, 109, 120f., 124, 130ff.
Maxwell's Demon 132
Maxwell 132
Meaning 2, 9, 27, 31, 72, 78f., 111, 115
Measurability 14, 26f., 97f.
Measurement result 56, 59
Measurement 55, 57f., 93
Memory content 44
Memory image 42f., 48f.
Memory 32, 42ff., 48f., 67, 75, 77, 82, 111f.
Memory 49, 71f., 77, 107f., 111ff., 126f., 129
messenger substance 65
Meta target value 29
Metabolism 17ff., 29
Metabolism 18
metaphysical 83
Mind 42, 48f., 52, 75, 77
Molecule 81
Monism 11, 99
Monkeys 48
Monod 134
Motif 50f.
Movement 14, 31, 84, 121
Mozart 73f.
Muscle elongation 42
Muscle 29
Natu 2, 132, 134ff.
Natural Law 114
Natural science 2, 7, 11, 13, 32, 45, 103, 105, 122, 134ff.,

145
Nature image 85
Nature 2, 7, 9, 11ff., 15, 17, 19f., 23ff., 32f., 37ff., 45f., 50, 60, 68, 82f., 85, 88, 96, 98f., 103, 105f., 110, 113f., 122f., 130, 132, 134ff., 145
necessities of life 21, 27, 33
Need 14, 25, 134
Need 17ff., 23, 25f., 36, 46, 50, 114
Nerve cell 30
Nerves 30, 41, 49, 60, 70, 72, 112
Nest 43f.
Neumann 134
non corporeal 124
non-local 8, 36, 55f., 58, 80f., 83, 85ff., 93ff., 102f., 105ff., 122f.
Normal value 28f.
Nutrition 18ff., 30, 43
Odour sensor 43
Operations 135
Order 84
orderly information 90f.
Organization 18, 38, 40, 79, 101, 103
organized 15, 18, 33, 38, 94, 102, 104
Overlay 57, 93
ovum 102f.
Oxygen 13ff., 82
Oysters 21f.
Parallelism 99, 116
Particle accelerator 89
particle physics 89, 121
Particle 56f., 59, 81f., 84ff., 89, 92f., 121f.
particles, for example the common oscillation state (polarization) of two entangled light particles. 56
Penrose 134
Performance 34, 42f., 48ff., 54, 61f., 68, 73, 75, 83f.,

139

124ff.
Personality 47, 52, 64, 78, 95, 104, 107ff., 111, 117, 120, 122, 127, 130
Phenomenon 2, 7f., 55, 58f., 81, 95, 121, 129, 135
Philosopher 133f.
Philosophy 7, 31f., 46, 68, 110, 133f.
Photon 55, 57f., 89, 92, 106
physical 48, 50f., 54, 61, 66f., 79, 103, 123f., 127
physical 8, 13ff., 24ff., 32, 36, 55, 79, 81, 84f., 87ff., 94f., 102, 105ff., 117, 120ff.
Physics 2, 8, 13, 23, 55ff., 81, 83ff., 90, 93, 103, 121ff., 129, 133ff.
Plant 11f., 22, 31, 37f., 41, 82, 101, 103f.
Polarization 56ff.
Power transmission 30
Practicality 17, 20, 25ff., 33, 39, 94
Pressure 42
Prigogine 134
Prime number 75
Principle 11, 27f., 30, 32f., 37, 41, 80, 99f., 102ff., 108, 128
Probability 24, 60, 64, 93, 115f.
Probe 23
Process 16f., 27f., 34ff., 40ff., 44, 46, 48f., 52, 78f., 82, 85ff., 90, 92, 95, 99ff., 105f., 117, 120ff.
Proof 135
propagation 46, 108
Property 12, 15, 20, 57, 66, 75, 84ff., 87f., 93, 95, 98, 107, 129
Protein 15, 19, 35
Protoplasm 15, 18, 20, 38, 101f.
Psychology 55, 59, 61, 69f., 76, 134

Purpose 13, 16f., 20, 25ff., 33, 39f., 60, 94, 109f., 112, 115
Quant 133f., 136
Quantum Correlation 136
Quantum entanglement 30, 55f., 59, 80ff., 87, 93, 95, 103, 121, 129
Quantum entanglement 8
quantum field 84
quantum information 133
Quantum Leap 121
quantum mechanics 57
Quantum physics 7, 58, 81, 134, 136
Quantum Theory 134
Quantum 7f., 30, 55ff., 80ff., 84, 87, 93ff., 103, 117, 121f., 129, 133f., 136, 145
Radiation 92
Random information 90f.
Random 43, 45, 64, 90f., 121f., 133f.
real 57, 83, 88, 91, 93, 121, 134
Reality 68, 88, 93, 134
Reality 83ff., 88
Reason 7, 9, 49, 68, 77
receptor 29
Recognize 44
reflection 134
reflex 134
Regeneration 17, 21ff., 33
Regulation process 103
Regulation system 28, 30ff., 40ff., 47f., 51ff., 75f., 78ff., 94f., 97ff., 105, 112, 128
Regulation 28, 30ff., 40ff., 47f., 51ff., 75f., 78ff., 94f., 97ff., 112, 116, 128
Relationship 31, 52, 68, 81, 129
Relativity 132f.
Religion 134, 136
Remote action 8, 55, 58f., 80f., 85f., 93, 103
Reproduction 29

Richet 60
Robot 135
Room 2, 84, 135
salamander 39
Schleich 67, 145
Schopenhauer 50
Schrenck notzing 135
Schrödinger's cat 57
Schrödinger 135
Science 132, 134, 136
scientific 8, 12, 31f., 34, 80, 83, 122
Scourge 29
Sea 12, 22, 43, 68
Secret 29
Secret 9f., 36f., 96, 101, 104f., 107f., 122f., 127
secretion 29
Self-confidence 46ff.
Self-employment 53, 62, 100, 109, 123
Self-organisation 133
Sensation 41ff., 54, 104
sense of sight 84
sensory cell 51
Sensory experience 54, 76, 111, 113, 116, 128
Sensory Organ 54, 79ff., 110, 126
Sensory stimulation 49
Shaft 35
Shannon 90, 135
Signal 29
size 84
Skills 21f., 42, 70f., 77f., 107f., 111f., 116, 125f., 128f.
Sleep 53, 61ff., 65ff., 69ff., 73f., 124f.
somnambul 70f., 73f.
Somnambulism 69f., 72
Soul 31f., 34, 36, 53, 55, 63, 73, 101, 128f., 132ff.
Sound wave 42
Space and time 8, 55, 68, 128, 132
Space-time 2, 55, 78, 87, 93,

140

95, 103, 135
Sparrow136
spatial expansion85ff., 93f.
Spatiality84
Spectrum132ff., 136
Speed of light81, 86, 93f., 103
Speed81, 93f., 103
Spirit31, 34, 48f., 53, 68, 74, 108, 111f., 132ff.
spooky8, 55, 58f., 81
Starfish22
State90
steamy12, 14f.
Stimulation51
stimulus processing101, 109
stimulus16, 38, 41f., 44ff., 49, 54, 61, 68, 101, 109
Subconscious76ff., 82
Substance information89, 91f., 94f., 106, 117
sugar19f., 26
Suggestion54, 66, 76f., 107, 111
Superposition state57
Supervisors' Association135
System13, 17f., 24, 29f., 32, 34f., 42f., 47, 56, 68, 80, 89ff., 109, 120, 135f.

Szilard136
tactile sense84
Target7, 10, 18, 28f., 46, 95, 108ff.
Tartini63
teleportation132
Temperature zero point90, 121
Temperature12, 14, 42, 90, 121
Term83ff.
Theory of relativity68, 87, 89, 132f.
Theory55f., 59, 79, 81, 83, 95, 122
Thinking7, 9, 26, 48f., 51f., 54, 62, 78, 92, 106ff., 110f., 122, 126ff., 130
Thirst50
This side36, 56, 87, 93ff., 117, 120f., 127f.
Thought Experiment57
Tier11f., 21f., 31, 38f., 41ff., 65, 75f., 101, 103f., 133
Time85, 136
Tipler136
Total information90f.
Touch8, 20, 42, 81

Transcendence133
Traum61, 63ff., 68, 74, 82
Unfolding105, 115, 125f., 129
Universe121, 132
Vacuum87f., 121
Venter16
Vitalism24
Waking state61, 67, 70, 76
Wallace60
Water splitting136
Water11ff., 25, 39, 49f., 66, 101, 136
Weight13
Wheat sacks133, 136
Wilber136
Will30, 50ff., 69
Wool48, 52, 54, 80, 97, 107, 109, 126f., 130
Work12f., 20, 23, 26, 35, 62, 117
World Puzzle133
World85, 133ff.
Wound21ff., 57, 59, 74, 92
Yeast fungus20f.
You prel63
Zeilinger58, 106, 132, 136
zymase20

NATURWISSENSCHAFT, PHYSIK UND ASTRONOMIE

– **Äquivalenz von Information und Energie.** Von: K.-D. Sedlacek

– **Das Gesetz im Zufall:** Wie sich verborgene Gesetzlichkeit manifestiert. Von: Moritz Cantor u. K.-D. Sedlacek (Hrsg.)

– **Die Transzendenz der Realität :** Spuren einer allumfassenden transzendenten Realität jenseits von Raum und Zeit. Von: K.-D. Sedlacek

– **Einsteins Relativitätstheorie ganz ohne Mathematik.** Spezielle und allgemeine Relativitätstheorie. Von: Prof. Dr. Paul Kirchberger u. K.-D. Sedlacek (Hrsg.)

– **Freizeitvergnügen Sternenhimmel mit bloßem Auge:** Wie man Sternbilder auffindet ohne Instrumente. Von: Prof. Dr. Paul Kirchberger u. K.-D. Sedlacek (Hrsg.)

– **Phänomen Naturgesetze:** Das Geheimnis hinter den Erscheinungen der Welt. Von: K.-D. Sedlacek

– **Supervereinigung:** Wie aus nichts alles entsteht. Von: K.-D. Sedlacek

– **Die Natur psycho-physikalischer Phänomene.** Erforschung telekinetischer Vorgänge. Von: Schrenck-Notzing, A. u. Klaus D Sedlacek (Hrsg.)

– **Giganten der Physik.** Die Top10-Physiker der Menschheitsgeschichte. Von: Klaus-Dieter Sedlacek (Hrsg.)

– **Der allmächtige Informatiker:** Das Mysterium des Universums. Von Sir James Jeans u. K.-D. Sedlacek (Hrsg.)

– **Der verborgene Mechanismus des Weltgeschehens:** Neue Erkenntnisse über die Gestalten biotechnischer Systeme der Welt. Von: Dr. h. c. Raoul Francé u. K.-D. Sedlacek

– **Der erdgeschichtliche Klimawandel:** Den wahren Ursachen von Klimaschwankungen auf der Spur. Von Wilhelm Bölsche u. K.-D. Sedlacek (Hrsg.)

– **Wege zur physikalischen Erkenntnis.** Meine wissenschaftlichen Selbstbiographie, Reden und Vorträge. Von **Max Planck** u. K.-D. Sedlacek (Hrsg.)

– **Leonardo da Vinci:** Seine naturwissenschaftlichen Studien und genialen Erfindungen. Von Hermann Grothe u. K.-D. Sedlacek (Hrsg.).

– **The philosophy of physical science.** By Sir Arthur Eddington.

– **The nature of the physical world.** By Sir Arthur Eddington.

– **Leben in der Warmzeit der Erde.** Aus den Urtagen vor dem heutigen Klimawandel. Von Wilhelm Bölsche und K.-D. Sedlacek (Hrsg.

– **Treibhauseffekt und Klimawandel:** Energiewende, ja bitte, aber nicht wegen CO_2. Von Klaus-Dieter Sedlacek (Hrsg.)

– **Über die Gewissheit von Vorhersagen:** Wahrscheinlichkeiten bestimmen ohne Formelballast. Von Klaus-Dieter Sedlacek (Hrsg.)

CHEMIE

– **Der Stein der Weisen:** Wie die Alchemie zur Chemie wurde. Von: Wilhelm Ostwald et. al. u. K.-D. Sedlacek (Hrsg.)

– **Durchblick Chemie:** Praktische Grundlagen und Einführung in die anorganische, organische und Biochemie. Von: Prof. Dr. Lassar-Cohn, Prof. Dr. W. Löb, K.-D. Sedlacek

NATUR- UND PHILOSOPHIE

– **Die letzten Ursachen.** Das Buch der Naturerkenntnis. Von: K.-D. Sedlacek

– **Gebundener Wille:** Wie frei ist menschlicher Wille tatsächlich? Von: K.-D. Sedlacek, G.F. Lipps et. al.

– **Jenseits der Erscheinungen:** Erkenbarkeit und Realität der Quantennatur. Von: Prof. Dr. M. Schlick u. K.-D. Sedlacek (Hrsg.)

– **Kleines Wörterbuch der Natur-Philosophie:** 1200 Begriffe, die man kennen sollte, kurz und prägnant. Von: K.-D. Sedlacek

- **Naturphilosophie:** Das Wesen von Naturgesetzen und die Erklärung des Lebens. Von: Prof. Dr. M. Schlick u. K.-D. Sedlacek (Hrsg.)
- **Vereinbarkeit von Religion und Naturwissenschaft.** Von: Kurd Laßwitz u. K.-D. Sedlacek (Hrsg.)
- **Das Konzept des Guten.** Sinnliches Empfinden – Der Ursprung unserer Wertvorstellungen. Von: Klaus-Dieter Sedlacek (Hrsg.)
- **Ist echte Erkenntnis möglich?** Einführung in die Erkenntnistheorie. Von: Prof. Dr. Erich Becher u. K.-D. Sedlacek (Hrsg.)
- **Das individuelle Ich**: Was ist der Kern des Selbstbewusstseins? Von: Th. Lipps u. K.-D. Sedlacek (Hrsg.).
- **Persönlichkeit und Unsterblichkeit:** In welcher Form existiert ein Weiterleben nach dem zeitlichen Ende? Von: Wilhelm Ostwald u. K.-D. Sedlacek (Hrsg.)
- **Die idealistischen Grundwerte unserer Kultur.** Von Johannes M. Verweyen u. K.-D. Sedlacek (Hrsg.)
- **Was sind Wirklichkeiten?** Aufgedeckte Naturgeheimnisse. Von: Kurd Laßwitz u. K.-D. Sedlacek (Hrsg.)

BEWUSSTSEIN

- **Leben nach dem Leben:** Befreiung des Bewusstseins von den Fesseln der Zeit. Von: K.-D. Sedlacek
- **Quantenbewusstsein.** Von: N. Wrobel u. K.-D. Sedlacek
- **Synthetisches Bewusstsein.** Von: K.-D. Sedlacek
- **Unsterbliches Bewusstsein:** Raumzeit-Phänomene, Beweise und Visionen. Von: K.-D. Sedlacek

LEBEN UND MEDIZIN

- **Leben aus Quantenstaub.** Von: N. Wrobel u. K.-D. Sedlacek,
- **Was ist Krankheit?** Von: N. Wrobel u. K.-D. Sedlacek
- **Bewusstsein und Unsterblichkeit.** Von: C. L. Schleich u. K.-D. Sedlacek (Hrsg.)
- **Die Lebenskraft:** Wie Enzyme, Bewusstsein und quantenbiologische Effekte das Leben regulieren. Von: K.-D. Sedlacek u. N. Wrobel,
- **Die verborgene Ordnung des Weltsystems.** Neue Erkenntnisse über die schöpferischen Kräfte der Natur. Von: Dr. h. c. Raoul Francé u. K.-D. Sedlacek (Hrsg.)
- **Homöopathie und Praxis:** Naturheilkundliche alternative Medizin für den mündigen Patienten. Von: Dr. med. J. Voorhoeve u. K.-D. Sedlacek (Hrsg.)
- **Eine andere Sicht auf die Entstehung der sporadischen Form der Alzheimerkrankheit.** Von Norbert Wrobel u. K.-D. Sedlacek (Hrsg.)
- **Bleib beweglich und fit ohne Geräte.** Leichte ärztliche Zimmergymnastik für jedes Alter. Von Moritz Schreber.
- **Plötzlich gesund.** Medizinische Wunderheilungen und die Macht organische Leiden psychisch zu beeinflussen. Von Erwin Liek.

PSYCHOLOGIE

- **Gestalt-Psychologie:** Einführung in die neue Psychologie vom Begründer der Gestaltpsychologie. Von: Prof. Dr. Kurt Koffka u. K.-D. Sedlacek (Hrsg.)
- **Die ersten Spuren psychischer Erscheinungen:** Das psychische Leben von Mikroorganismen – Eine Studie in experimenteller Psychologie. Von Alfred Binet u. K.-D. Sedlacek (Übers.)
- **Allgemeine moderne Psychologie:** Systematische Einführung in die Wissenschaft psychischer Prozesse. Von August Messer u. K.-D. Sedlacek (Hrsg.).
- **Strahlende Kräfte durch positives Denken:** Die Wurzeln des Erfolgs und Wege zum Glück. Von Emil Peters u. K.-D. Sedlacek (Hrsg.)
- **Neue praktische Menschenkenntnis.** Ein Ratgeber zur Menschenbehandlung mit

zahlreichen Bildern und Beispielen. Von Johannes Maria Verweyen.

– Massenpsychologie am Beispiel Jan Bockelsons. Geschichte eines Massenwahns mit einer Einführung von Sigmund Freud. Von Friedrich Reck-Malleczewen u. K.-D. Sedlacek (Hrsg.)

BIOLOGIE

– Wie intelligent sind Pflanzen? Sensationelle Einblicke in die geheime Seite des pflanzlichen Wesens. Von Prof. Dr. phil. Adolf Wagner u. K.-D. Sedlacek

– Über Menschenaffen, Tierseele und Menschenseele: Intelligenzprüfungen an Hominiden. Von Wilhelm Bölsche et. al. und K.-D. Sedlacek (Hrsg.)

GESCHICHTE, VOR- U. FRÜHGESCHICHTE

– Die geheimnisvolle Kultur der alten Kelten. Von Druiden, Fürstensitzen und der Lebensart unserer frühgeschichtlichen Vorfahren. Von Georg Grupp u. K.-D. Sedlacek (Hrsg.)

– Der Alchemist Leonhard Thurneysser: Die Lebensgeschichte des Goldmachers von Berlin. Von Klaus-Dieter Sedlacek (Hrsg.)

– Es begann mit Feuerskraft. Das Werden des Menschen und seiner Kultur. Von Carl W. Neumann u. K.-D. Sedlacek (Hrsg.)

– Gefangen zwischen Eisschollen: Die dramatische Entdeckungsgeschichte der Antarktis. Von Klaus-Dieter Sedlacek (Hrsg.)

RATGEER

– Kultur erleben mit den Wohnmobil in Frankreich: Vierzig kulturelle Highlights, Park- und Übernachtungspätze sowie Navigationskoordinaten. Von Klaus-Dieter Sedlacek

– Kochbuch für ganze Kerle: Kräftige und Feinschmeckergerichte für Freizeit und Camping. Von K.-D. Sedlacek (Hrsg.)

– Der Weg zu Wohlstand und Reichtum: Goldene Regeln für den Aufbau einer selbständigen Existenz. Von P.T. Barnum u. K.-D. Sedlacek (Hrsg.)

– Die Kultur der Azteken: Mit einem Anhang Große Landesausstellung Baden-Württemberg „Azteken" im Lindenmuseum. Von William Prescott.

FORSCHUNGSREISEN U. ABENTEUER

– Meine erste Weltumseglung: Tagebuch einer epochalen Expedition. Von James Cook u. K.-D. Sedlacek (Hrsg.)

– Exotische Reise durch Persien: Abenteuerlicher Bericht aus einer fremdartigen Welt des 19ten Jahrhunderts. Von Pierre Loti u. K.-D. Sedlacek (Hrsg.)

– Mit der Beagle um die Welt: Bericht meiner Forschungsreise zum Galapagos-Archipel. Von Charles Darwin u. K.-D. Sedlacek (Hrsg.)

– Peking-Paris im Automobil: Die legendäre 16.000 km – Rallye 1907. Von Luigi Barzini u. K.-D. Sedlacek (Hrsg.)

Buchshop:

www.ingramcontent.com/pod-product-compliance
Lightning Source LLC
Chambersburg PA
CBHW031920240526
45464CB00021B/618